TALZOYA

Your Brain and Your Happiness

First published by Voltis Press 2023

Copyright © 2023 by Talzoya

Library of Congress Control Number: 2023930274

First edition

ISBN: 978-1-7353758-7-8

This book was professionally typeset on Reedsy.
Find out more at reedsy.com

Contents

Preface

One personality trait in humans is how sensitive and responsive we are to incentives and rewards, such as food, sex, money, praise, education, and professional achievements. This trait motivates us to pursue goals related to rewards. Research shows that the feelings of being elated and excited because we are moving toward achieving an important goal are biochemically based, though they can be modified by experience. Another personality trait is how sensitive and responsive we are to threats and punishments. The feelings of anxiety and discomfort that we experience in the face of uncertainty and danger are also controlled biochemically. Our brains differ not just in their tendency to focus our attention more on the desire to obtain the rewards that we do not currently have versus avoiding the dangers that may lie ahead, but also in how much pleasure we are able to derive from the rewards that we currently possess and how much pain we feel from the consequences of mistakes already made. These nervous system differences, in turn, critically affect our subjective well-being or ill-being, or what we call happiness.

Individuals whose brains are wired to be more sensitive to rewards than punishments have a temperamental predisposition to feel positive affect most of the time and to be happy, while those whose brains are naturally more responsive to punishments than to rewards are predisposed to negative affect and ill-being. From an evolutionary perspective, while varying traits on a continuum among the individuals within a group can be beneficial for the survival of the group as a whole, the resulting traits can have both fortunate or unfortunate effects on the subjective well-being of the individuals themselves. At the personal level, extreme variation in traits can result in lopsided temperaments, such as stable extraverts versus neurotic introverts, sometimes even giving rise to mental illnesses, such as mania and depression.

The action of genes on our emotions, thoughts, and behavior is pervasive. Even the likelihood of getting divorced has a strong heritable component! A 2018 research study by Jessica Salvatore, from the Virginia Commonwealth University, and her collaborators of a large sample of adoptees found consistent evidence that genetic factors contributed to the intergenerational transmission of divorce but weaker evidence for a rearing-environment effect on divorce. In their history of divorce, adoptees resembled their biological parents rather than their adoptive parents. The tendency to divorce is, in part, the result of individual differences in personality, of which the stable component is strongly heritable. It should not be surprising, therefore, that individual differences in happiness, too, stem to a large degree from genetic variation.

Interpersonal variation in temperament reflects underlying differences in brain structure and biochemistry. If happiness is the end goal of all human endeavor, then understanding how the brain regulates individual reaction to rewards and punishments, how it directs attention, and how it generates patterns of approach and avoidance can provide opportunities to increase subjective well-being and decrease human suffering. While personality traits establish our baseline levels of joy, meaning, contentment, and so on, this well-being set point is then pushed and pulled about as good and bad things happen to us. Which personality traits enhance happiness and which undermine it? Do certain life events have the power to permanently modify our happiness set point? What actions can we take to spend most of our lives in the upper half of our biological happiness range? How can we set and adjust life goals to maximize our well-being? Can money buy happiness? This book will attempt to answer these questions and more based on scientific evidence.

Acknowledgement

I want to thank my editor, Chris Gage, for his thorough work and professionalism.

I

The Biological Basis of Happiness: Why Some People Are Consistently Happier Than Others

1

Personality and Happiness

"Are these things good for any other reason except that they end in pleasure and get rid of and avert pain? Are you looking to any other standard but pleasure and pain when you call them good?" With these two questions directed at the renowned sophist Protagoras, Socrates famously identifies the pleasurable and the good along with the painful and the bad. The excess of pleasure over pain in one's daily life, which is akin to the psychological notion of *affect balance*—the difference between the amount of positive affect and the amount of negative affect at any point in time—together with an optimistic evaluation of one's overall satisfaction with life represent the core of what we call *happiness* or *subjective well-being*. The notion of happiness thus includes an affective component representing the tint of our emotional mental state or mood and a cognitive component based on an analytical assessment of the overall quality of our life.

Moods and emotions, which together are labeled *affect,* represent people's instinctive evaluations of the events that are currently occurring in their lives. While emotions are fleeting mental states, moods are typically more enduring. Both are aimed at focusing our attention and action toward events and things in our environment that are either beneficial or harmful to our survival and reproduction. While a relatively unimportant occurrence may trigger a momentary shift in our emotions, more important situations—those that call for sustained attention and effort—engender moods. Two dominant dimensions consistently emerge in studies of affective structure: positive affect and negative affect. Despite what their names may suggest, positive affect and negative affect do not represent the opposite ends of the same concept; instead, they are two independent factors that scientists typically measure separately. Positive affect reflects the extent to which a person feels joyful, optimistic, enthusiastic, active, engaged, and alert. High positive affect is a state of high energy, full concentration, and pleasurable engagement with the environment, whereas low positive affect is characterized by indifference toward one's environment, dullness, and lethargy. In contrast, negative affect is a general dimension of subjective distress, unpleasant engagement with the environment, active avoidance, and a variety of aversive mood states, including anger, contempt, disgust, guilt, shame, hopelessness, fear, and nervousness. Low negative affect is a state of calmness, contentment, and serenity. Both positive and negative affect are strong and independent predictors of a person's happiness. Affect and personality traits are so intimately tied that scientists often have difficulty distinguishing the items on a mood measure from those on a personality inventory. Both have substantial genetic underpinnings and are thus reflections of enduring dispositions.

In a comprehensive review article published in 1999, Ed Diener and his collaborators, from the University of Illinois at Urbana-Champaign, found that personality is one of the strongest and most consistent predictors of subjective well-being, meaning that people have a genetic predisposition to be happy or unhappy, which is presumably caused by inborn individual differences in the nervous system. Personality represents constitutional differences in reactivity and self-regulation among people. These differences stem from enduring traits that propel individuals to consistently think, feel, and behave in specific ways, and they are shaped by heredity, maturation, and experience. Individuals differ in how they react to changes in their environment, as reflected in the activation and deactivation patterns in their brains and endocrine systems. They also differ in how they allocate attentional resources and selectively engage in approach and avoidance behaviors with respect to rewards and punishments. These enduring patterns of reactivity and self-regulation collectively form the basis of personality.

The five-factor model of personality posits five broad dimensions used in common language to describe the human personality: neuroticism (being sensitive and nervous versus resilient and confident), extraversion (being outgoing and energetic versus solitary and reserved), openness to experience (being inventive and curious versus consistent and cautious), agreeableness (being friendly and compassionate versus critical and rational), and conscientiousness (being efficient and organized versus extravagant and careless). Research consistently shows that the traits of extraversion and neuroticism are particularly important with respect to subjective well-being.

Extraversion and introversion are the two ends of a single continuum. Extraverts (who score high on the extraversion scale) are often perceived as enthusiastic, full-of-energy, action-oriented individuals. They enjoy interacting with people and engaging in a variety of activities. In group settings, they are gregarious, assertive, talkative, and typically perceived as dominant. They are highly motivated by the pursuit of rewards, such as sex, money, career advancement, social standing, and praise. They take pleasure in activities that involve large social gatherings, such as parties, community activities, public demonstrations, and business or political associations. They also tend to work well in groups. Introverts (who score low on the extraversion scale), by contrast, have lower social engagement and energy levels than extraverts. They tend to appear reserved, low-key, deliberate, and socially aloof. Their lack of social involvement is not due to shyness; they just need less stimulation and more time alone than extraverts. Introversion is an innate preference, while shyness stems from distress. Introverts prefer solitary to social activities but do not necessarily fear social encounters like shy people do. They tend to enjoy a quiet and minimally stimulating external environment. They prefer to concentrate on a single activity at a time and like to observe situations before they participate. They tend to be slower, more deliberate, and more analytical in their actions.

Neuroticism and emotional stability also represent the two ends of a single continuum. Neurotics (who score high on the neuroticism scale) are emotionally reactive and vulnerable to stress. Because they are particularly sensitive to threats and punishments, they are more likely to interpret

5

ordinary situations as ominous. They can perceive minor frustrations as hopelessly difficult. They are more likely than average to be moody and to experience such feelings as anxiety, worry, guilt, shame, envy, jealousy, alienation, loneliness, and resentment. They are described as often being self-conscious and shy, complaining endlessly, and tending to have trouble controlling urges and delaying gratification. This may explain why they also tend to experience more negative life events and are especially susceptible to developing a number of psychopathologies, including anxiety disorder, depression, bipolar disorder, substance abuse disorders, eating disorders, and schizophrenia. Emotionally stable individuals (who score low on the neuroticism scale) tend to be calm, composed, balanced, and free from persistent negative feelings. They remain even-tempered, particularly in the face of challenges and threats. Emotional stability is similar to resilience because both terms involve being able to withstand hardship. Being emotionally stable does not mean that you are always content. It just means that when you do find yourself in emotionally challenging situations, you respond in a reasonable way while maintaining a sense of composure. Emotionally stable people are generally free of psychopathologies and can work effectively under stressful conditions.

In 1980, Paul Costa and Robert McCrae, from the National Institutes of Health, proposed a model relating extraversion to positive affect and neuroticism to negative affect. According to their model, someone with a temperamental disposition for extraversion should be sociable, energetic, and engaged, which will result in positive affect and an overall feeling of satisfaction with life. A highly neurotic person will be prone to anxiety, hostility, impulsivity, and psychosomatic complaints, which will result in negative affect and an overall feeling of dissatisfaction with life. These two components are subjectively balanced out by the individual to arrive at a net sense of subjective well-being, which may be measured as morale, life satisfaction, hopefulness, or simply happiness. It should be noted that the personality factors of neuroticism and extraversion are independent from one another, meaning that a person can be extraverted and neurotic, extraverted and emotionally stable, introverted and neurotic, or introverted and emotionally stable. Emotionally stable introverts and neurotic extraverts may have similar levels of life satisfaction or happiness, but they achieve this result in utterly different ways. Emotionally stable introverts are seldom depressed but just as seldom elated and may report an average level of happiness, all things considered. Neurotic extraverts are prone to extremes of both hopelessness and optimism depending on life circumstances and reach average overall satisfaction only because there is as much positive affect as negative affect in their lives. Individuals with the lowest probability for overall psychological well-being are neurotic introverts who not only do not feel much enthusiasm or energy in life, but are also especially prone to anxiety, irritability, and pain. Those with the highest probability of subjective well-being are stable extraverts with their low capacity for negative affect and high capacity for positive affect. It is important to note that the vast majority of people are in the midrange of both the extraversion and neuroticism scales, implying that they tend to be relatively contented most of the time. What all this means is that, to a large extent, an individual's capacity for happiness or unhappiness is an enduring, trait-like characteristic given that adult personality is found to be strongly stable over a person's lifetime.

According to an article published in 1994 by Robert McCrae and Paul Costa, the five factors of personality (neuroticism, extraversion, openness to experience, agreeableness, and conscientiousness) have been measured in a score of longitudinal studies that followed the same set of individuals over periods of more than two decades using a variety of samples, instruments, and designs. These studies produced the following remarkably consistent results:

1. The mean level of personality traits change with development, but they reach final adult levels at about age thirty. Between age twenty and thirty, both men and women become somewhat less emotional and thrill-seeking and somewhat more cooperative and self-disciplined—changes we might interpret as evidence of increased maturity. After age thirty, there are few and subtle changes, of which the most consistent is a small

decline in activity level with advancing age. Except among individuals with dementia, stereotypes that depict older people as being withdrawn, depressed, or rigid are unfounded.

2. Individual differences in personality traits, which show at least some continuity from early childhood on, are also essentially fixed by age thirty. Personality traits are found to be quite stable even over intervals as long as thirty years, although there is some decline in stability magnitude with increasing retest interval, meaning that the traits can have modest long-term changes.

3. Stability appears to characterize all five of the major domains of personality—neuroticism, extraversion, openness to experience, agreeableness, and conscientiousness. This finding suggests that an adult's personality profile as a whole will change little over time.

4. Generalizations about stability apply to virtually everyone. Men and women, healthy and sick people, blacks and whites all show the same pattern. When asked, most adults will say that their personality has not changed much in adulthood, but even those who claim to have had major changes show little objective evidence of change on repeated administrations of personality questionnaires. Important exceptions to this generalization include people suffering from dementia and certain categories of psychiatric patients who respond to therapy, but no moderators of stability among healthy adults have yet been identified.

Here is how McCrae and Costa commented on these findings: ". . . people undoubtedly do change across the life span. Marriages end in divorce, professional careers are started in midlife, fashions and attitudes change with the times. Yet often the same traits can be seen in new guises: Intellectual curiosity merely shifts from one field to another, avid gardening replaces avid tennis, one abusive relationship is followed by another. Many of these changes are best regarded as variations on the 'uniform tune' played by individuals' enduring dispositions. . . . Those individuals who are anxious, quarrelsome, and lazy might be understandably distressed to think that they are likely to stay that way, but those who are imaginative, affectionate, and carefree at age thirty should be glad hear that they will probably be imaginative, affectionate, and carefree at age ninety." The stability of personality is what makes life predictable to some extent. We can make vocational and retirement lifestyle choices with some confidence that our current interests and enthusiasms will not desert us any time soon. We can choose friends and mates with whom we are likely to remain compatible. We can know which coworkers we can depend on, and which we cannot. There are such things as spontaneity and impulse in human life, but they appear to be relatively stable traits.

The strongest evidence for a temperamental predisposition to experience relatively consistent levels of subjective well-being across the lifespan comes from behavior-genetic studies of heritability. Heritability studies estimate the amount of variance in subjective well-being scores among individuals that can be explained by genes. In 1988, Auke Tellegen, from the University of Minnesota, and his collaborators examined monozygotic and dizygotic twins who were reared together and others who were reared apart. Monozygotic twins (also called identical twins) result from the fertilization of a single egg by a single sperm, with the fertilized egg then splitting into two. Identical twins share the same genomes and are always of the same sex. In contrast, dizygotic twins (also called fraternal twins) result from the fertilization of two separate eggs with two different sperm during the same pregnancy. They share half of their genomes, just like any other siblings. Fraternal twins may not be of the same sex or have similar appearances.

The scientists assessed these twins' personality using the Multidimensional Personality Questionnaire. Comparing identical twins reared together to those reared apart (and cross-checking with data on fraternal twins reared together and apart) allows the scientists to separate out genetic influences on personality factors from the influence of the shared family environment. They found that monozygotic twins who grew up in different homes were more similar to each other than were dizygotic twins who were raised together or apart. Furthermore, dizygotic twins who were raised in the same family were not much more similar to each other than were dizygotic twins who were raised apart. Tellegen and his collaborators estimated that genes account for about 40 percent of the variance in positive emotionality (or extraversion) and 55 percent of the variance in negative emotionality (or neuroticism), whereas shared family environment accounted for 22 percent and 2 percent of the variance in positive emotionality and negative emotionality, respectively. This means that the variance in adult happiness at any given point in time is determined about equally by genetic factors and by the effects of experiences that are, for the most part, unique to each individual. Family environment appears to contribute almost nothing to negative emotionality, while it accounts for about one-fifth of positive emotionality. Positive emotionality is most clearly an affective and interactive dimension. High positive emotionality is characterized by active engagement with one's environment, whereas low positive emotionality is characterized by weak engagement or disengagement. Because of its inherently interactive and communicative character, it seems plausible that a person's positive emotionality trait level is particularly responsive to and reflective of the surrounding social climate, including the prevailing and more or less engaging climate of the rearing environment. The finding that the common environment generally plays a very modest role in the determination of many personality traits runs counter to the traditional belief, influential among psychologists, that personality similarity is profoundly enhanced by a shared family environment. Environment does seem to carry substantial weight in the determination of personality—it appears to account for at least half the variance—but that environment is one for which twin pairs are correlated close to zero. Rearing dizygotic twins in the same family environment does not lead to greater similarity between them. When it comes to personality, these scientists seem to see environmental effects that operate almost randomly with respect to the sorts of variables that psychologists (and other people) have traditionally deemed important in personality development.

David Lykken and Auke Tellegen retested a subset of the same twins after intervals of four-and-a-half and ten years and published their results in a 1996 paper. They found that the heritability of the stable component of personality and subjective well-being approaches 80 percent. This means that, although 40 percent to 55 percent of the variation among individuals in current or short-term levels of subjective well-being can be explained by genes, 80 percent of the variation in long-term subjective well-being is due to genetics. The researchers also found that, over the long run, educational attainment and socioeconomic status accounted for about 3 percent of the variance in subjective well-being, and income for about 2 percent, but marital status only accounted for less than 1 percent of the variance, while religiousness did not account for much of the variance in happiness at all. These events would be expected to have stronger effects on short-term levels of subjective well-being, meaning right around the time that we finish college, start a new career, get a significant pay raise, get married, or decide to join a religious group. As time passes by, we habituate to our new state of affairs, and our subjective well-being slowly drifts right back to our stable level, which is determined by our genes.

Here is how Lykken and Tellegen commented on their results: "No one doubts that making the team, being promoted at work, or winning the lottery tends to bring about an increment in happiness, just as flunking out, being laid off, or a disastrous investment would be likely to diminish one's feelings of well-being. . . . [H]owever, the effects of these events appear to be transitory fluctuations about a stable temperamental set point or trait that is characteristic of the individual. Middle-aged people whose life circumstances have stabilized seem to be equally contented regardless of their social status or their income. The reported well-being of one's identical twin, either now or ten years earlier, is a far better predictor of one's self-rated happiness than is one's own educational achievement, income, or status." They add: "We know that when people with bipolar mood disorder are depressed, they tend to avoid intimate encounters or new experiences and tend to brood upon depressing thoughts rather than concentrating on their work. Then, when their mood swings toward elation, these same people tend to do the things that happy people do. There is undoubtedly a James-Lange feedback effect: Dysfunctional behavior exacerbates depression, whereas the things happy people do enhance their happiness. We argue, however, that the impetus is

11

greater from mood to behavior than in the reverse direction." Based on their long-term heritability estimate for happiness of 80 percent, it could be said that it is almost as hard to change one's happiness as it is to change one's height. Such pessimism is, however, unwarranted. Even the long-term estimate of the heritability of happiness leaves room for environmental influence, and the estimate of one's current level of subjective well-being leaves about half of our happiness level up to environmental effects. Thus, one can focus on happiness at a specific period in life and conclude that heritability has a moderate influence, or one can focus only on people's average happiness over the long term (for instance, a decade) and conclude that heritability has a substantial effect.

Studies such as the ones by Tellegen and collaborators are the basis for the *set point theory of happiness*. The set point theory of happiness posits that our average level of subjective well-being is determined primarily by heredity and by personality traits ingrained in us early in life, so it remains relatively constant throughout our lives. Our level of happiness may change transiently in response to life events, but then it almost always returns to its baseline level as we habituate to those events and their consequences over time. Inborn temperament not only sets the average mood around which our emotions fluctuate up and down, but it also sets an emotional range, unique to each individual, within which those fluctuations can take place.

Let's imagine a person who scores extremely high on both the extraversion and the neuroticism traits, perhaps scoring a nine on a one-to-ten scale on both factors. This person may have an average set point for happiness, but they can feel extremes of both elation and sadness depending on their life circumstances. They would thus have a much wider emotional range than an average person and be much more reactive than average to environmental cues. An event that may leave an average person contented may render them exhilarated, while a relatively minor frustration may make them deeply despondent. They would presumably need a particularly robust nervous system capable of handling such wide fluctuations, absent of which they are at risk of developing a psychopathology—in particular, bipolar disorder. A person with bipolar disorder can experience periods of relative mood stability, but they can also fluctuate between mania   an enduring state of extremely high energy, activity, and mood—and depression—a persistent state of unusually low energy, activity, and mood. Someone scoring very low on extraversion (say, a two on a one-to-ten scale) and very high on neuroticism (perhaps a nine on a one-to-ten scale) will have a low set point for happiness; they may be melancholic or mildly depressed most of the time. While they can never feel much in the way of contentment in life, they can feel extremes of anguish and despondency at times. Their emotional range is about half as wide as that of the previous person and shifted toward the low end of the happiness scale. One can intuitively guess that this person is probably at risk for a whole host of mental disorders, including unipolar depression, anxiety, and schizophrenia (a severe condition similar to having anxiety, depression, and constant psychosis or detachment from reality, all at the same time). In contrast, a person who scores very high on extraversion and very low on neuroticism will have a sunny disposition as a set point. While they can never get too upset about anything in life, they can feel extremes of exhilaration at times. A dysfunctional nervous system may make them prone to mania but not to depression. Some men exhibit this type of "bipolar disorder," and their condition should probably be renamed "unipolar mania." Someone scoring very low on extraversion and very low on neuroticism (perhaps a two on a one-to-ten scale for both factors) may have an average set point for happiness but has the narrowest possible emotional range. Such a person would not be very reactive at all to environmental change, no matter how good or bad the events that may befall them. They would generally exhibit low activity levels, and the flatness of their affect may lead others to judge them as being cold, aloof, and even insensitive. One can speculate that this type of extreme temperament may represent a predisposition for narcolepsy, a chronic neurological disorder that affects the brain's ability to control sleep-wake cycles and makes people feel very tired throughout much of the day.

In 2006, Bruce Headey, from the University of Melbourne in Australia, proposed a slight revision to the set point theory of happiness based on the analysis of data from the German Socioeconomic Panel (SOEP). At the time that Headey made his claims, the West Germany segment of the SOEP was the longest running panel in the world to collect data on subjective well-being. It began in 1984 with a sample of 12,541 respondents. As of the publication date of the paper, interviews had been conducted annually ever since. During the interviews, respondents were asked to make summary judgments about their subjective well-being, defined as overall life satisfaction (which was rated on a scale of zero to ten), moods, or other mental states. Three-item scales were also included in surveys for each of the personality factors of neuroticism, extraversion, openness to experience, agreeableness, and conscientiousness. Headey analyzed survey data over the twenty-year span between 1985 and 2004. The result was that, although subjective well-being changed little over long time periods for most people, small but nontrivial minorities experienced substantial and apparently permanent upward or downward changes in happiness. A core finding from this analysis was that the people most likely to record large long-term changes in life satisfaction were those who scored high on one or more of the three personality traits of extraversion, neuroticism, and openness to experience. In a sense, these people tend to "roll the dice" more often than others in life, and thus have a higher than average probability of recording long-term changes in life satisfaction.

In light of his findings, Headey derived the following four conclusions as a modification of the set point theory of happiness: (1) Even in the long term, the large majority of people do not deviate much from their own equilibrium level or set point for life satisfaction. They are particularly unlikely to deviate if they score near the mean or below on extraversion, neuroticism, and openness to experience; (2) high levels of extraversion and openness to experience, combined with low neuroticism, are associated with high upside risk of favorable life events and substantial gains in long-term life satisfaction; (3) high levels of neuroticism and openness to experience, combined with low extraversion, are associated with high downside risk of adverse life events and substantial decline in long-term life satisfaction; and (4) high levels of extraversion, high levels of neuroticism, and high levels of openness to experience increase both upside and downside risk of both major life events and long-term change in life satisfaction.

The key to reaching these conclusions is the inclusion of the trait *openness to experience*, besides extraversion and neuroticism, in evaluating happiness levels. Openness to experience is a general appreciation for art, emotion, adventure, unusual ideas, imagination, curiosity, and variety of experience. People who score high on openness to experience are intellectually curious, open to emotion, sensitive to beauty, willing to try new things, and more likely to hold unconventional beliefs. Those who score low on openness to experience tend to be more pragmatic, traditional, restrained, closed-minded, and even dogmatic. They tend to like habits, and they very rarely deviate from them. Scoring high on openness to experience makes people take more risks or "roll the dice" more often, making them more likely to experience extremely positive life events if they are also high scorers on extraversion and low scorers on neuroticism (such as starting a new business and attaining fame and fortune as a result of it) or extremely negative life events if they also score high on neuroticism while scoring low on extraversion (such as becoming hopelessly addicted to heroin in an attempt to quiet down one's anxiety and ending up losing one's job, spouse, and home as a result of it). Scoring high on all three factors of openness to experience, extraversion, and neuroticism opens the person to the possibility of experiencing extremes of both positive and negative life events, which can increase or decrease happiness depending on the scale and valence of the events. Such a person could win tens of millions of dollars in a lottery, buy their dream home and retire, feel extremely happy, then develop a taste for gambling and alcohol, squander all of their winnings, and end up feeling miserable.

Do the remaining two personality factors of agreeableness and conscientiousness have an impact on personal happiness at all? Answering this question was one of the objectives of a 1991 study by Robert McCrae and Paul Costa. Conscientiousness is a tendency to display self-control, act dutifully, and strive for achievement against widely established measures or outside expectations. It is related to the way in which people control, regulate, and direct their impulses. People who score high on conscientiousness tend to be focused, self-monitoring, goal-setting, hard-working, efficient, and reliable. Low conscientiousness is associated with flexibility and spontaneity, but it can also appear as sloppiness and lack of reliability. People who lack conscientiousness can seem lazy, unprofessional, disorganized, irresponsible, undependable, and fickle. The agreeableness trait reflects individual differences in general concern for social harmony. Those who score high on agreeableness value getting along with others, have an optimistic view of human nature, and are generally caring, considerate, kind, generous, trusting, trustworthy, and helpful. People who score low on agreeableness tend to place their own self-interest above getting along with others. They are generally unconcerned with others' well-being and may be competitive, argumentative, untrusting, untrustworthy, and selfish. Sometimes their skepticism about other people's motives causes them to be suspicious, unfriendly, and uncooperative. McCrae and Costa found that high scores on both agreeableness and conscientiousness lead to higher life quality, result in more positive affect and less negative affect, and thus higher levels of subjective well-being. These traits exert an instrumental effect on happiness, in contrast to extraversion and neuroticism, which directly produce positive and negative affect, respectively. Agreeable individuals are warm, generous, and loving; conscientious people are efficient, competent, and hard-working. The interpersonal bonds that agreeableness fosters and the achievements and accomplishments that conscientiousness promotes may contribute to greater quality of life and, instrumentally or indirectly, to higher levels of happiness.

J. A. Gray's theory of personality provides the biological reason why extraverts and neurotics have a natural susceptibility to experience positive and negative affect, respectively. Gray contended that there are two neurologically based motivational systems responsible for many of the observed behavioral and emotional differences between extraverts and neurotics. One of these neuronal systems, the behavioral activation system, is thought to regulate behavior in the presence of signals of reward, such as sex, food, safety, achievement, aggression, and social attachment. Rewards and opportunities trigger mental processes that help us pursue and obtain goals related to these rewards. The behavioral activation system triggers affective states such as hope, excitement, joy, pride, and liveliness. The other system, the behavioral inhibition system, is thought to regulate behavior in the presence of signals of punishment, threat, and uncertainty. The behavioral inhibition system involves affective states such as anxiety, heightened vigilance, inspection of punishment contingencies, avoidance, and response inhibition.

Gray suggested that individuals differ in the relative strengths of these two signal-sensitivity systems, with extraversion relating to a strong sensitivity to signals of reward, and neuroticism relating to a strong sensitivity to signals of punishment. Extraverts are more sensitive to rewards, and this sensitivity manifests itself in the form of greater pleasant affect when exposed to rewarding stimuli. Higher positive affect then motivates individuals to approach rewarding stimuli. Because social situations tend to be more fun and rewarding than nonsocial situations, extraverts' elevated positive affect and sensitivity to rewards lead to increased social behavior. Extraverts are not necessarily caring; the propensity to care for the well-being of others is controlled by the personality factor of agreeableness, which is independent of extraversion. Extraverts can score high or low on agreeableness. An extravert who scores low on agreeableness is said to exhibit agentic extraversion, which involves incentive motivation and is expressed as a tendency toward assertiveness, persistence, and achievement. It is all about using social ties as a way of obtaining rewards, social rank, and ultimately power. An extravert who also scores high on agreeableness exhibits affiliative extraversion, which involves the positive emotion of social warmth and is expressed as a tendency toward amicability, gregariousness, care, and affection.

Neurotics, on the other hand, are more sensitive to punishments, and this sensitivity manifests itself in the form of greater unpleasant affect when exposed to threatening stimuli. Higher negative affect then motivates individuals to inhibit their behavior in order to avoid the punishing stimuli. Because social situations can sometimes be felt as threatening (for instance, when one is in the presence of a more powerful person), neurotics can feel incentivized to avoid others in order to escape any harm that they might inflict upon them, especially if they also score low on agreeableness. It is thus consistent with Gray's theory to hypothesize that positive affect and negative affect are the state manifestations of reward-signal sensitivity and punishment-signal sensitivity, respectively. When exposed to signals of reward, one experiences positive affect, and when exposed to signals of punishment, one experiences negative affect, and some people are naturally wired to be more sensitive to rewards than an average person, while others are more sensitive to punishments than average.

One can easily intuit how variability in characteristics such as "fearfulness" or "cheerfulness" might provide some adaptive benefit to individuals living together in groups. It allows for functional specialization with respect to punishments and rewards. Those more attuned to threats might fulfill the role of sentinels, alerting others to looming dangers, while those who are more sensitive to rewards may help the group survive by assuming the role of providers. This functional specialization allows individuals of each type to be more effective at their specific task by delegating the responsibility for the other task to their peers. It is, therefore, not surprising that individual differences in emotion processing are particularly pronounced among social species such as primates and are probably most extreme among humans. At the individual level, the extraordinary heterogeneity in how different individuals respond to the same emotionally provocative challenge plays a crucial role in shaping variations in well-being. Affective style refers to consistent individual differences in emotional reactivity and regulation. Many of the parameters of affective style, such as the threshold to respond to an emotional event, magnitude of response, latency to peak of response, and recovery function, are features that are often opaque to individual consciousness, though they may influence the subjective experience of emotion.

In studying the neurobiological basis of affective style, Richard Davidson, from the University of Wisconsin at Madison, has uncovered the importance of the recovery function following negative events for vulnerability to certain forms of psychopathology, as well as for resilience. Failure to recover rapidly following a negative event can be a crucial ingredient of vulnerability to both anxiety and mood disorders, particularly when such a style is combined with frequent exposure to negative events over a sustained period of time.

By contrast, the capacity for rapid recovery following negative events may represent an important ingredient of resilience, which is defined as the maintenance of high levels of positive affect and well-being in the face of significant adversity. It is not that resilient individuals never experience negative affect, but rather that the negative affect does not persist. Such individuals are able to profit from the information provided by the negative affect, and their capacity for "meaning making" in response to such events may be part and parcel of their ability to show rapid decrements in their body's stress response following exposure to a negative or stressful event.

In studying the brain circuitry responsible for the processing of emotions, Davidson and his research collaborators have uncovered important functional differences between the left and right sides of the human prefrontal cortex. The prefrontal cortex is the cerebral cortex covering the front part of the frontal lobe, the area of the brain directly behind the forehead. This brain region has been implicated in planning complex cognitive behavior, personality expression, decision making, and moderating social behavior. Studies comparing the mood of patients with unilateral left-sided or right-sided frontal brain damage have found a greater incidence of depressive symptoms following left-sided damage. The general interpretation of these studies is that depressive symptoms are increased following damage to the left-sided prefrontal cortex because this brain territory participates in certain forms of positive affect and, when damaged, leads to deficits in the capacity to experience positive affect, a hallmark feature of depression. Studies in which scientists image the brain of normal people and those with anxiety disorders generally find that increases in right-sided activation in various sectors of the prefrontal cortex are associated with increased negative affect. Ten-month-old babies who cry in response to maternal separation are more likely to have less left-sided and greater right-sided prefrontal activation during brain imaging studies compared with those infants who do not cry in response to this challenge. Baseline measures of asymmetric prefrontal activation predict reports of well-being among individuals in their late fifties. According to Andrew Tomarken, Richard Davidson, and their collaborators, there is now substantial evidence from studies using a variety of subject populations (for example, neurobiological, psychiatric, and normal) and research methods that increased activation of the left prefrontal cortex is associated with heightened positive affect, decreased negative affect, or both. Conversely, increased relative activation of the right prefrontal cortex is linked to heightened negative affect, decreased positive affect, or both. These findings support the idea that individual differences in patterns of prefrontal cortex activation mark some aspect of vulnerability to positive and negative emotion elicitors, which is then expressed in terms of higher reward versus punishment sensitivity and, ultimately, in the personality factors of extraversion and neuroticism. These brain activation patterns are already present in newborns, produce observable differences in mood and behavior starting in adolescence, and remain relatively stable throughout adulthood.

The prefrontal cortex is the latest evolutionary addition to the human brain, and human emotion certainly predates its development. Ample research has been devoted to understanding which subcortical brain regions are involved in the processing of positive and negative emotions and how those regions interact with the prefrontal cortex to produce the observed variety of individual responses to emotionally provocative challenges. An important finding of this body of research is that the human brain appears to have different systems for the processing of immediate versus anticipated rewards and threats. Individual differences in the capacity to feel *desire*, the strong longing to have something one does not currently have, versus *dread*, the fearful anticipation of punishment, have deep implications for variations in personality and, hence, subjective well-being.

2

Desire and Dread

According to Daniel Lieberman and Michael Long, John Douglas Pettigrew, an Australian neuroscientist, was the first to clarify how the human brain creates a three-dimensional map of the world while working on the idea that fruit bats are our distant cousins. Although our space is often perceived as unitary, an ever-growing body of neuropsychological evidence demonstrates that the brain actually contains a modular representation of space, some cortical regions being involved in the processing of the *extrapersonal space*, which is the space that is far away from us and that cannot be directly acted upon by the body, while other cortical regions appear to process the *peripersonal space*, which is the space that directly surrounds us and with which we can directly interact. Peripersonal space includes everything within arm's reach, things we can control by using our hands. This is the world of what is real, right now. Extrapersonal space refers to everything else, things we cannot touch unless we move beyond our arm's reach. This is the realm of possibility and imagination. Since moving from one place to another takes time, any interaction in the extrapersonal space must occur in the future, meaning that distance is linked to time. Acquiring something out of our immediate reach may also require some planning. This is the defining characteristic of things in the extrapersonal space; getting them often happens *reflectively* or via thought and requires motivation, effort, time, and, typically, planning. By contrast, anything in the peripersonal space can be experienced right here and right now, without much effort or planning, often *reflexively* or out of instinct. From an evolutionary standpoint, the food, water, or shelter that you do not currently have are critically different from the food, water, or shelter that you do have. And what is true of the immediate versus future rewards is also true of the immediate versus future dangers. Becoming penniless in retirement is a danger lurking in the distant future for a person in their mid-twenties, and avoiding this fate requires careful planning and sustained motivation and effort over a long period of time. In contrast, the car that is about to hit us unless we run very fast as we cross the road is a real danger menacing us right now. This distinction between immediate versus delayed rewards and punishments is so fundamental that the human brain has evolved two separate systems to process the peripersonal and extrapersonal spaces.

Anticipated rewards trigger the feeling of desire and the motivation to pursue them, while anticipated punishments create the feeling of dread and the motivation to avoid them. The pursuit of rewards requires elevated energy, attentiveness to benefit-related cues, pleasant engagement with the external environment—whether it be social or physical—and a general state of activation of both body and mind. The avoidance of dangers, on the other hand, involves heightened levels of fear, lower energy, unpleasant engagement or disengagement with the external environment—and this includes both people and things—and a general state of inhibition of both body and mind. Danger denotes a situation that is perceived to be both inescapable and very painful. This type of situation calls for behavioral inhibition. A danger, in this sense, is different from a challenge, which represents a risky but potentially rewarding situation. A challenge calls for behavioral activation. A brain chemical called dopamine and its associated circuitry are in charge of managing everything that lies in the extrapersonal space, things that are distant in space and time. Acting together, they generate both positive reward motivation toward appetitive stimuli and negative fear motivation toward aversive stimuli. This desire-and-dread circuitry represents the entirety of what J. A. Gray referred to as behavioral activation and behavioral inhibition systems in explaining the neurobiology of personality, in particular the personality factors of extraversion and neuroticism. What he saw as two neurologically distinct systems turns out to be a single system—our brain's dopamine circuitry—that can be tuned in the direction of either desire and activation or dread and inhibition based on both environmental cues and the biological wiring of the system itself.

The dopamine desire-and-dread circuitry is powerful. It can focus attention or scatter it, motivate action or inhibit it, and either thrill or terrify us. Whether it generates excitement or fear, dopamine's role is the same: a relentless focus on enhancing our future. The pursuit of rewards that one perceives to be attainable triggers a surge in dopamine, which brings about energy, enthusiasm, and hope. Life feels good. Extraverts, who are especially sensitive to rewards, spend their lives pursuing this thrill of anticipation, the feeling that life is about to get better. The dopamine circuitry makes us passionately want things, but appreciating a reward that we have earned is processed by an entirely separate circuitry—one we will discuss in the next chapter—which is in charge of processing the peripersonal space. Once a desired thing becomes ours, the dopamine circuitry quiets down because it does not process experience in the real world, only in the imaginary future. In a way, desire dopamine promises delights that it is in no position to deliver. For many, this is a letdown. Pursuing the object of our desire often feels much better than enjoying it once in our possession. Similarly, avoiding threats that one perceives to be unsurmountable also triggers a surge in dopamine, this time bringing about a feeling of anxiety, pain, lethargy, and hopelessness. Life feels awful. Neurotics, who are especially sensitive to threats, spend their lives constantly worrying about the future, dreading an impending doom. When an event they feared comes to pass and reality sets in, it is often a relief; imagination can produce doomsday scenarios that rarely live up to dopamine's hype. So they, too, can often feel cheated by their own brain.

Dopamine works by imparting a crucial property called salience to environmental cues: *appetitive or incentive salience* to rewarding stimuli and *aversive or fearful salience* to threatening ones. Salience refers to the extent to which things are important, prominent, novel, or conspicuous compared with their surroundings. One kind of salience is the quality of being unusual.

If an element seems to jump out from its environment, it is salient. If it blends into the background and takes a while to find, it is not. Familiarity typically decreases the salience of novel stimuli over time so that you notice them less or not at all.

For instance, when you wear contact lenses for the very first time, you are aware of their presence throughout the day. Your brain detects these strange new objects laying on top of your cornea and makes you pay attention; perhaps you should be concerned and take action. As you continue wearing the contact lenses day in and day out, your brain recognizes them as helpful objects that are familiar and not worth noticing anymore, and you no longer are conscious of their presence unless they start bothering you in some way. Another kind of salience is emotional value. Things are salient if they have the potential to affect our future for better or for worse or if they are important to us personally because of their potential to impact our well-being. Rewarding and threatening stimuli are salient because they may affect our chances of survival and reproduction; they trigger a surge of dopamine that makes us pay attention and become excited or fearful.

Because we have limited perceptual and cognitive resources, dopamine focuses our attention toward sensory data that it judges to be pertinent or salient to the detriment of all other data. When the brain excludes unimportant details from consciousness, it is said to engage in latent inhibition. The word *latent* describes things that are hidden, and the brain inhibits latent things or conceals them from consciousness because they are unimportant. Our brain automatically uses salience attribution and latent inhibition to highlight the things that it judges to be important, making them rise to consciousness, and to hide those that it judges to be unimportant, making them absent from conscious awareness. As a result of this process, two people who live in the very same physical household may perceive the world in utterly different ways because they have different thresholds for salience attribution and latent inhibition, and this causes them to have different personalities. The brain component called the hippocampus, which processes memories, is part of the dopamine network. It helps with the assessment of salience and context by using past memories to filter new incoming stimuli and placing those that are most important into long-term memory. Research shows that it is a brain region that suffers damage early on in Alzheimer's disease, one of the effects of which is altered salience attribution and, along with it, altered personality.

Without our brain's capacity to assign salience and engage in latent inhibition (meaning without dopamine) it would be utterly impossible for us to construct a mental representation of the world—our personal models about how things work and fit together. Model building uses salience attribution and latent inhibition because models only contain the elements that the model builder believes are essential. Models make the world easier to comprehend and, later, to imagine a variety of ways it might be acted upon to benefit us.

Unbeknownst to us, our brains start building rudimentary models of the world in childhood, which are then updated and modified continuously as we grow into adults and learn new things. As we gain experience with the world, we develop improved models, and this is the basis of wisdom. Models allow us to simplify our conception of the world but, even more important, they allow us to abstract, to take specific experiences and use them to craft broad, general rules about how things work. Our prefrontal cortex can retrieve several models relevant to our specific situation from long-term memory, hold them in working memory while it analyzes which one fits the situation at hand best, and then decide what action to take or choice to make based on that information. According to Daniel Gilbert, author of *Stumbling on Happiness*, the greatest achievement of the human brain is its ability to imagine objects and episodes that do not exist in the realm of the real—an ability made possible by our mental models of the world—and it is this ability that allows us to think about the future in a process called mental time travel. Mental time travel is a powerful tool of the dopamine system. It allows us to experience a possible future that does not presently exist as if we were there. The emotions that this experience elicits are the basis of the choices that we make in the present moment. Mental time travel is the mechanism we use for every single conscious choice in life, from deciding which pair of socks to wear for the day to choosing a vacation destination to electing a profession to pursue. How well our personal models fit the real world is very important. If our models are poor, we will make bad predictions about the future, which will result in bad decisions.

In the late eighteenth century, Immanuel Kant triggered a philosophical earthquake when he came up with the idea that our perceptions are not the result of a purely physiological process by which our senses somehow transmit an exact image of the world into our brains, but rather, they are the result of a psychological process that combines what we see, hear, touch, smell, and taste with what we already think, feel, know, want, and believe, and then uses this combination of sensory information, preexisting knowledge, and emotions to construct a perception of reality. So the world as we know it is a construction, a manufactured article of the mind, to which the mind contributes as much as the environmental stimuli themselves. Maturity is part and parcel of the realization that perceptions are merely points of view, that what we see is not necessarily what there is, and that two people may have very different perceptions about the same thing or circumstance. Emotionally stable extraverts, with their sunny disposition, optimism, and energy are likely to build much more cheerful and brighter models of the world than neurotic introverts who are besieged by negative affect and bestowed with a natural inclination to imagine the worst possible outcome of any action or event. Where one sees a pool of opportunities, the other one only sees obstacles, and where one sees cause for celebration, the other sees reason to brood. These two archetypes appear to live side by side in very different worlds. Understandably, the more extreme a person scores on the extraversion and neuroticism scales, the more distorted their internal models are likely to be compared to those of a neurotypical person. These divergent views of the world—which tend to remain relatively stable throughout adulthood, except in cases of mental illness or dementia—critically affect a person's subjective well-being. Indeed, they reflect our happiness set point and range.

Extreme emotional reactivity, whether in the positive or negative direction, represents a vulnerability to psychopathology. All that is needed to light up the fire is the right environmental circumstance. A moderate upward deviation from average in positive affect, yielding mildly elevated mood and activity for an extended time, represents a state of hypomania. Another notch up from there with a touch of psychosis may warrant a diagnosis of mania. Continuing upward in affect and adding a bit more psychosis, we get mania with psychotic features, which usually comes with a dose of anxiety, shifting the mood from elated to irritable. With a lot more psychosis, the diagnosis changes to schizoaffective disorder. If we push further along in the same direction, the mood becomes even more irritable, the anxiety worse, and the psychosis constant and severe and we now have a case of schizophrenia. Similarly, a moderate downward deviation from average in negative affect leads to melancholia or mild depression. If we continue further down, we get depression, perhaps mixed with some anxiety. When the mood and activity levels are further depressed and a touch of psychosis is added, the affected person may have anxiety and depression with psychotic features. If we go even further down in affect and add even more psychosis, the diagnosis becomes schizoaffective disorder. At the extreme end of the negative affect scale with continuous and severe psychosis, we get schizophrenia. This description represents the full psychosis spectrum. It seems clear that neither a continuous state of exhilaration nor extended melancholia are desirable. Perhaps what we call happiness in common language refers to a baseline of mild euphoria or calm contentment, along with a capacity to feel moderate levels of excitement or disappointment as warranted by circumstances and the ability for a relatively rapid recovery or return to our baseline.

It appears that psychosis is a necessary consequence of the extreme distortion of both salience attribution and latent inhibition, which happens when we deviate too much from the average in affective tone, a direct consequence of too much dopamine in the brain and a nervous system that is not robust enough to handle it. A psychotic person's mental representation of the world becomes too detached from the actual world, often in ways that seem bizarre. In a 2005 article, Shitij Kapur, Romina Mizrahi, and Ming Li, from the Centre for Addiction and Mental Health in Toronto, Canada, proposed that psychosis arises from an aberrant assignment of novelty and salience to objects and associations. Antipsychotics block dopamine receptors and decrease dopamine transmission, which leads to the attenuation of aberrant novelty and salience. While the hallmarks of established psychosis are delusions (fixed, false beliefs) and hallucinations (aberrant perceptions), patients usually experience months of a prodromal period that predates the expression of frank psychosis.

Kapur and collaborators postulate that, during the prodrome, there is a context-independent or context-inappropriate firing of dopamine cells and dopamine release. This produces a perplexing sense of novelty and salience in patients—a state well captured in the following patients' accounts of the prodromal period: "My senses were sharpened. I became fascinated by the little insignificant things around me"; or "I noticed things I had never noticed before." Patients continue to accumulate these experiences of altered novelty and salience without a clear explanation for them. There is a gradually increasing sense of perplexity, confusion, and alterations in mood and behavior until it all crystallizes into a delusion. Delusions represent the explanations that the patient constructs in an effort to make sense of the aberrant salience experiences. Since the individual constructs the delusions, they are imbued with the psychodynamic themes and cultural context of that individual. This may explain how the same dopamine system dysfunction convinces a patient in an African village that he is the victim of the black magic by an evil shaman, while the student in Toronto is convinced that the Royal Canadian Mounted Police is using the internet to monitor her.

Imagine that the patient in the African village visited the local shaman to get relief from his prodromal symptoms out of a lack of access to mental health services or simply ignorance. During the visit, the shaman touches the patient's forehead and utters a few wishes of good health. Suppose that the very next day, the patient's symptoms suddenly and coincidentally take a turn for the worse; he feels foggy and forgetful, has a feeling of terror he cannot shake, has lost the motivation to perform even the smallest daily routines, and experiences extreme difficulty getting out of bed because of an inexplicably painful sensation in his legs. As the days and weeks pass and the patient continues to suffer, he starts to see the visit to the shaman as particularly salient until he becomes fixated on the idea that the shaman must have cast an evil spell on him while touching his forehead. Why would the shaman touch his forehead otherwise? Why not any other body part? Surely, it was the shaman who messed up his head. There can be no other explanation for the painful symptoms. The patient rehashes these thoughts in his head until they become unshakeable beliefs or delusions. While distorted salience attribution can lead to delusions, distorted latent inhibition can lead to hallucinations.

Let's imagine how a tactile hallucination may form. Our brain normally conceals much of the regular functioning of our body from our conscious awareness. For instance, our internal organs are constantly wobbling as we move around. The awareness of this constant wobbling is typically suppressed because it would be too distracting from regular functioning. A dysfunction of the dopamine system may disrupt latent inhibition, resulting in a schizophrenia patient's awareness of the movements of their internal organs. As a result of this distressing experience, the patient may become paranoid about the possibility of losing their internal organs, perhaps as a result of tearing away due to excessive wobbling. Although this sensation is rooted in reality, because it is suppressed from consciousness for most of us, a neurotypical person will interpret the patient's account of it as a bizarre, imagined sensation, a hallucination. Now, let's think about how auditory and visual hallucinations may develop. Both our imagination and senses use the very same processing areas of the brain in producing our experiences, whether real or imagined. Our brain normally has a "reality first" policy; if imagination and reality are competing for use of the same systems at the same time, a healthy brain will suppress imagination in favor of reality. When latent inhibition is disrupted, this "reality first" policy is broken, and imagination can seep into conscientiousness as if real. So our internal voice (which all of us use without normally being confused about its nature) starts to feel as real as the voice of another person, and the characters that we sometimes imagine for ourselves project in front of us as if real. Visual and auditory hallucinations are the most common type of hallucinations reported by schizophrenia patients. One can only imagine the level of confusion, fear, and distress that an afflicted person can feel when this happens to them for the first time.

Before discussing our brain's dopamine circuitry in more detail, a brief description of how the human nervous system works is in order. Our nervous system is made of the central nervous system, consisting of our brain and spinal cord, and the peripheral nervous system, which includes the rest of our nerves and sense organs. Its function is to regulate information, meaning to receive, transmit, and process it. The nerve cell or neuron is the building block of the nervous system. It consists of a cell body equipped with many short tentacles called dendrites, which bring information in, and a long, cordlike extension called an axon, which transmits information out to the next cell. The nerves that run throughout our body consist of groups of axons banded together. A synapse is the minute space separating the axon of a signal-passing neuron (the presynaptic neuron) and one of the dendrites of the target neuron (the postsynaptic neuron). An electrical signal travels through the length of one cell to stimulate the release of a packet of chemical messengers called monoamine neurotransmitters when it reaches the end of its axon. Dopamine is one such neurotransmitter. These neurotransmitters are transported out of the axon into the synapse and stimulate the receptors on the edge of the next cell. This, in turn, induces the receiving cell to send an electrical signal along its length. Reuptake is the reabsorption by a presynaptic neuron of monoamine neurotransmitters that it has just secreted and released. It happens when excessive concentrations of neurotransmitters are detected in the synaptic plasma. Monoamine transporters are protein structures that lie just outside of the presynaptic cleft and serve to transport or mop up the neurotransmitters from the synapse back into the neuron that emitted them. Transporters and receptors are commonly associated with drugs used to treat mental disorders, as well as recreational drugs, a line that can become quite blurred at times.

Dopamine is produced in two small brain structures called ventral tegmental area (VTA) and substantia nigra. Both of these structures are located in the midbrain, the most forward portion of the brain stem, which connects the cerebrum and the spinal cord. VTA dopamine is in charge of motivation, meaning the assignment of either incentive salience or fearful salience to environmental stimuli, while dopamine from the substantia nigra controls either initiation or inhibition of action, meaning the initial muscle movements that we must either make or prevent in order to either approach or avoid a stimulus based on the valence of the VTA-initiated motivation.

Parkinson's disease causes the neurons in the substantia nigra to slowly die. Once roughly 90 percent of the substantia nigra is destroyed, levels of dopamine in the brain's motor circuitry become so low that the patient starts to experience significant symptoms of the disease: tremors, rigidity, slowness of movement, and difficulty with walking, among other ailments. The typical treatment for Parkinson's disease is the administration of a drug called L-dopa, also known as levodopa, which works by increasing the levels of dopamine in the brain. The substantia nigra dopamine projects into the prefrontal cortex, and also into a part of the brain called the striatum (located just above the brain stem and below the cortex), and these three structures communicate through dopamine signals to control initiation of movement. The VTA dopamine also projects into the prefrontal cortex and, additionally, into a small structure right next to the striatum called the nucleus accumbens. While the VTA, prefrontal cortex, and nucleus accumbens together play a crucial role in generating motivation, a small nucleus called the amygdala (located at the terminal end of the striatum) exerts a modulatory influence on the nucleus accumbens to shift the valence of the motivation from positive and approach-oriented to negative and avoidance-oriented. The hippocampus, which is also located in the vicinity of the striatum and involved in memory, also receives dopamine signals from the VTA and is in charge of the long-term storage of salience associations. It is the VTA dopamine system that becomes dysfunctional in the brains of people with active schizophrenia. Given the nature of the roles played by the brain areas that are part of this system, it is not surprising that schizophrenia patients suffer from memory problems (stemming from dysfunction of the hippocampus), difficulty concentrating (due to dysfunction of certain parts of the prefrontal cortex), extreme fear (stemming from the amygdala), and distorted salience (due to dysfunction of the nucleus accumbens). There is too much dopamine circulating in this circuitry during a schizophrenia episode, confounding its operations. Treatment of schizophrenia typically includes an antipsychotic medicine along with other drugs. Antipsychotics work by blocking a type of dopamine receptors called D2 receptors, which prevents excessive intake of dopamine by target neurons in the nucleus accumbens. This alleviates the distortion of salience attribution and causes a decline in psychosis.

What I referred to as our brain's desire-and-dread circuitry at the beginning of this chapter is the brain's VTA dopamine pathway. The switching of the salience of stimuli from desired to dreaded happens in the nucleus accumbens, a central target of the VTA dopamine. The neurons of the nucleus accumbens, in addition to having a high concentration of dopamine receptors, also present with glutamate receptors. Glutamate is the most abundant excitatory neurotransmitter in the vertebrate nervous system, and various brain areas use it to communicate with one another.

In a 2008 experiment, Alexis Faure and collaborators, from the laboratory of Kent Berridge at the University of Michigan, selectively blocked glutamate receptors in the shell of the nucleus accumbens of rats. They found that blocking glutamate receptors in the rostral shell increased appetitive behavior such as eating, whereas blocking glutamate receptors in the caudal shell increased fearful reactions such as distress vocalizations and defensive treading. In neuroscience jargon, rostral refers to a location that is closest to the forehead and caudal, a location closest to the back of the head. As the blockade of the glutamate receptors gradually moves from rostral to caudal (or from front to back) in the nucleus accumbens shell, the behavior of the rats slowly shifts from appetitive to fearful, reminiscent of the shifting of sounds from low-pitched to high-pitched when a pianist moves her fingers from left to right on the keyboard. The scientists also found that dopamine helps to modulate the action of glutamate in assigning salience. The generation of appetitive behavior required not only the blockade of rostral glutamate receptors but also the activation of the D1 dopamine receptors. In contrast, the generation of fearful behavior required the blockade of the caudal glutamate receptors, as well as the activation of both the D1 and the D2 receptors by dopamine. The amygdala (working with the prefrontal cortex) is in charge of assessing whether something we see warrants vigilance or not, and it communicates this information to the nucleus accumbens via glutamate signals. So the amygdala is like the pianist moving her fingers on the shell of the nucleus accumbens, which works like a piano's keyboard. With some help from the VTA dopamine, this action imparts environmental stimuli with a tinge ranging from extremely attractive to extremely frightening. Since the activation of the D2 dopamine receptors is necessary for aversive salience, it should come as no surprise that the blockade of the D2 receptors by antipsychotics is what puts an end to the fearful paranoia seen in schizophrenic psychosis.

A neurotypical person's brain seamlessly assigns varying degrees of appetitive or aversive salience to environmental cues in direct proportion to what their circumstances warrant. In comparison, the brain of a person who scores very high on extraversion can sometimes get stuck in the appetitive salience mode in response to a positive event in their life, such as falling in love. When this mental state persists for an extended time period, the person may sometimes be slowly transitioning into a manic episode, in which they experience elevated mood (either euphoric or irritable), flight of ideas, pressured speech, increased energy, decreased need and desire for sleep, and hyperactivity. When the brain is locked in this incentive salience (or desire) mode, the VTA dopamine preferentially activates the left prefrontal cortex, and we experience positive affect along with behavioral activation. If the same person also scores extremely high on neuroticism, their brain can sometimes get stuck in the aversive salience mode in response to an adverse event in their life, such as getting divorced. When this mental state persists for an extended time period, the person may be slowly transitioning into a depressive episode, in which they experience dysphoric mood, low motivation, low energy, disturbed sleep or appetite, and reduced activity. When the brain is locked in this fearful salience (or dread) mode, the VTA dopamine preferentially engages the right prefrontal cortex, and we experience negative affect along with behavioral inhibition. Mood and activity seem to always work in a lockstep fashion, with elevated mood implying elevated activity and vice versa, and depressed mood implying depressed activity and vice versa. A brain equipped with a hyperactive dopamine circuitry resembles a high-performance car in a couple of ways. It is built for performance. With it, you can achieve the kind of endurance and energy needed to climb Mount Everest or solve a complex mathematical problem, assuming that you also possess the abilities required to accomplish these tasks. But push it a bit too far and, just like a high-performance car driven too hard, it can break down. Being neurotypical may make you generally happy, but it also makes you rather ordinary. Having a dopaminergic personality can make you creative, energetic, and persevering, but it can also make you vulnerable to the misery of mental illness. Genius sometimes results from aberrant salience or distorted latent inhibition, and happiness and achievement do not always go hand in hand.

The case of a sixty-five-year-old female patient in Germany presenting with advanced Parkinson's disease provides a vivid illustration of the tight connection between mood and movement. According to Dr. Jan Herzog and the team of doctors who treated her, by the time she first checked herself into her local hospital's neurology unit the patient had a fourteen-year history of Parkinson's disease, had been treated with L-dopa for ten years, and had needed surgical removal of part of her right thalamus for severe rigidity and tremor of the left side of her body.

The thalamus is our brain's information relay station. All information from our body's senses (except smell) must be processed through our thalamus before being sent to our brain's cerebral cortex for interpretation. The thalamus also relays information related to movement. Surgically removing part of the thalamus is an established therapy that provides immediate relief with uncontrolled movements that can occur in patients taking the drug L-dopa for a long time. The surgery is performed on the side opposite that of the worst tremors. Bilateral procedures are poorly tolerated because of increased complications and risk, including vision and speech problems, so surgeries are typically performed on one side of the brain only. The surgery gave the patient relief for about three years, after which she started experiencing increasingly distressing levels of tremors on both sides of her body. The neurosurgeons in charge of her case, Dr. Jan Herzog and his colleagues, decided to try deep brain stimulation of the subthalamic nucleus to provide her with additional relief. Both the thalamus and the subthalamic nucleus are in the vicinity of both the substantia nigra and the striatum, which, together with the prefrontal cortex represent the dopamine pathway that controls initiation of movement. The subthalamic nucleus is involved in modulating this motor circuitry via glutamate signals. The procedure consists of surgically implanting electrodes next to the subthalamic nucleus on each side of the patient's brain. Those electrodes can then be turned on by the neurosurgeon postoperatively, and the voltage level on them can be adjusted as needed. The neurosurgeons started the deep brain stimulation by turning on the electrodes in the brain of this patient six days after her surgery. This was followed by a rapid and marked improvement of motor functions associated with dyskinesia. Because she had some residual tremors, they also continued the L-dopa at a reduced dose. Neither psychotic episodes nor benign hallucinations induced by L-dopa had been reported by the patient or her caregivers before surgery. The patient did not present any symptoms of a psychotic or affective disorder, but the patient's mother had a history of bipolar affective disorder, indicating a genetic predisposition to affective disorders. Seven days after switching on the neurostimulator, however, a remarkable mood change took place. In parallel with motor improvement, the patient's mood was elevated to a degree that was abnormal for her. She was excessively talkative, and it was not possible to interrupt her while she was speaking. This indicated a state of hypomania. Gradually, over the next three weeks, her hypomania turned into a manic episode with psychotic symptoms. The patient's mood was euphoric and her speech more rapid, her thoughts were easily distracted, flights of ideas appeared, associations loosened, and her ability to concentrate faded. Orientation and memory were intact. She lost normal social inhibitions, was in love with two neurologists, and tried to embrace and kiss people. She was hyperactive and restless; she left the clinic several times without permission and engaged in unrestrained buying of clothing. Because of her disorganized behavior, she messed up her room, and she occupied her neighbor's bed. The patient alleged that important events were forthcoming and that she had to settle her affairs. Her family wanted to take her credit card from her to protect her from financial ruin. Her judgment was impaired, but she had little insight into her disorder and had low frustration tolerance. She was suspicious, tense, and hostile, and had the delusion that her sons were conspiring against her; she said that they tried to get her money by threat of force. There was a decreased need for sleep, and the patient felt rested after only three to four hours.

Attempts to decrease stimulation parameters or medication led to worsening of her motor state. Stimulation arrest led to a reappearance of severe tremors. Moreover, it was associated with a rapid deterioration of her mood with anhedonic feelings, but without any improvement in mania. Because of its unfavorable impact both on the motor and emotional state of the patient, stimulation arrest was not tolerated for more than half an hour. Her doctors started treating her with clozapine, an antipsychotic drug. Psychotic symptoms disappeared, but now mixed features with symptoms of both mania and depression were present at the same time. The patient was often emotionally labile, switching from laughter and euphoria to crying and depression in minutes. Periods of hyperactivity were followed by phases of complete fatigue, deep sadness, and grumbling that her situation had never been worse. Gradually, depressive symptoms tended to predominate. Because of the onset of mixed features with symptoms of mania and depression at the same time, with a gradual predomination of depressive symptoms turning into a depressive episode, her doctors diagnosed an organic bipolar affective disorder. For this reason, they reduced the dosage of her antipsychotic and added a mood stabilizer. With this combination, the affective disorder remitted within the next three months. They stopped the mood stabilizing treatment after fourteen months but continued the low-dose antipsychotic. In the following three years, she did not experience further manic or depressive episodes, and her motor symptoms were well controlled with continuous brain stimulation showing an improvement of 59 percent in comparison with the baseline. This case serves to illustrate the tight connection between mood and activity levels given that deep brain stimulation of the subthalamic nucleus seems to change not only motor behavior but also a person's emotional state in direct proportion to it. Our baseline mood and activity levels are reflections of our hard-wired reactivity toward rewards and punishments, and this reactivity appears to not only be a fundamental feature of human personality but also a strong determinant of our happiness set point.

Dopamine is not only implicated in determining the personality traits of extraversion and neuroticism through its action on the brain's D1 and D2 dopamine receptors, but it has also been determined to influence the variation in levels of novelty seeking among people via its interaction with the brain's D4 dopamine receptors. Individuals who score higher than average in novelty seeking are characterized as impulsive, exploratory, fickle, excitable, quick-tempered, and extravagant, whereas those who score lower than average tend to be reflective, rigid, loyal, stoic, slow-tempered, and frugal. Novelty seeking is most closely related to the personality trait of openness to experience in the five-factor model of personality. The gene that codes for the dopamine receptor D4, called DRD4, presents with several variations called alleles, which depend on the number of times that a certain sequence of DNA repeats itself in the gene, with the four- and seven-repeat alleles being the most common. The variations in the variable number of tandem repeats of DRD4 cause people to produce dopamine receptors of different sizes, a feature that is linked to the ability of the receptor to bind to dopamine-like molecules. In 1996, Richard Ebstein, from the Sarah Herzog Memorial Hospital in Jerusalem, Israel, and his research collaborators showed that higher than average novelty seeking scores in a group of 124 unrelated Israeli subjects were significantly associated with having the seven-repeat allele of the DRD4 gene. The association of high scores on novelty seeking and the long allele of DRD4 was independent of ethnicity, age, or sex of the subjects. These findings are consistent with experimental research on animals that has shown the role of dopamine receptors in exploratory behavior. For example, research on mice has shown that the injection of dopamine agonists into their brains markedly increases the frequency and duration of spontaneous exploratory activity and also facilitates the initiation, speed, and vigor of locomotion.

Another interesting feature of the DRD4 polymorphism is its great variation among human populations. East Asians have a low proportion of long alleles (typically 1 percent or fewer alleles with seven repeats), whereas South American Indians have a high proportion of them (up to 78 percent alleles with seven repeats). This knowledge inspired Chuansheng Chen, from the University of California at Irvine, and collaborators to surmise that migration may be a key natural selection factor that accounts for the global variation of DRD4 gene. Their analysis showed that populations that remained near their origins showed a lower proportion of long alleles of DRD4 than those that migrated farther away, a finding that remained consistent across all major six migration routes they considered.

For the longest route of migration (from northeastern Asia to Americas), South Americans have the largest proportion of long alleles (69 percent), followed by the one Central American group they considered (42 percent); whereas North Americans have the lowest proportion of long alleles (32 percent). Moreover, all American groups have more long alleles than northern and eastern Asians (for instance, the Yakut, Japanese, Chinese, and Taiwanese averaged 5 percent long alleles). The latter societies are presumed to share common ancestors with American Indians. Similarly, Jews who had migrated a longer distance eastward to Rome and Germany also showed a higher proportion of long alleles than those who migrated a shorter distance southward to Ethiopia and Yemen. Among the four groups of Africans they considered, the Bantu, who lived in South Africa and had migrated a long distance from Cameroon, showed a higher proportion of long alleles. Finally, among Indo-Europeans, the Sardinians, who live geographically closest to the origin of their language family, had zero percent long alleles, whereas the average for other European groups was 20 percent. Within each region, the societies with the lowest percentage of long alleles (for instance, the Pueblos in North America and the Quechuans in South America) were traditionally sedentary, whereas the other societies in the region were nomadic. It was estimated that the proportion of long alleles increased by 4.3 percentage points for each 1,000 miles of macro-migration.

The researchers surmise that groups migrated not because of the founding fathers' genetic makeup, but due to other causes, such as war and the depletion of natural resources. The eventual differences in the allele frequencies of

DRD4 between those who migrated and those who did not were a result of natural selection (or deselection) of the gene over the following millennia. Otherwise stated, those who migrated originally had the same proportion of long and short alleles of DRD4 as those who remained in place. Those among the migrants who possessed the long allele had a much higher inclination to explore their new environment than those possessing the short allele. This may have led them to discover and reap the benefits of the many rewards that the new environment offered more often than those with the short allele, leading them to be chosen as reproductive partners more often. Over time, this led to an overrepresentation of the long allele of DRD4 among the migrants. The farther the population migrated from its original location, the more exploratory opportunities presented themselves and the more dominant the long, seven-repeat allele became. Thus, it can be argued reasonably that exploratory behaviors are adaptive in migratory societies because they allowed for more successful exploitation of resources in the particular environment migration entails—usually harsh, frequently changing, and always providing a multitude of novel stimuli and ongoing challenges to survival. Conversely, sedentary populations obtain resources not by exploring new environments but by developing intensive methods for using limited amounts of land. Within these societies, novelty-seeking and exploratory behaviors would have serious social costs and would be selected against. Therefore, the fit between the behavioral consequences of long alleles of DRD4 in migratory societies ensures the selection of that gene and the lack of fit in sedentary societies ensures the deselection of that gene. The findings of this study also raise the possibility that some contemporary disorders, such as attention deficit hyperactivity disorder (ADHD), which has been linked to long alleles of the DRD4 gene, may have been a by-product of human adaptation to the migration process. The once-adaptive components of those disorders (such as high level of activity, attention diversion) became problematic in highly structured modern societies.

We mentioned earlier how scoring high on both the personality traits of openness to experience and neuroticism along with low scores on extraversion can make individuals likely to "roll the dice" more often, increasing their chances of experiencing negative life events. Indeed, contrary to those who score average on both openness to experience and neuroticism, they are more likely to take disadvantageous risks, such as abusing opiate drugs in an effort to tame excessive anxiety. It is a general research finding that high scores on the neuroticism scale represent a biological vulnerability to psychopathology and unhappiness. The tendency of neurotic individuals to focus on the negative aspects of life may be mediated by the action of the glutamate signals from their hyperreactive amygdala on the nucleus accumbens, tilting it more often in the direction of dread than desire. The useful role that the amygdala normally plays in helping us to avoid dangerous or disadvantageous situations can become problematic when it is much more reactive than average, which may be the case for those with high scores on the neuroticism scale.

The role of the amygdala in helping to generate aversive or fearful salience by inhibiting actions with potentially deleterious outcomes in a neurotypical brain has been shown in an experiment for which results were published in 2010 by Benedetto De Martino, Colin Camerer, and Ralph Adolphs, from the California Institute of Technology. Their study tested the hypothesis that the amygdala mediates loss aversion, an idea motivated by a large amount of extant literature implicating this brain structure in processing fear and threat, as well as in anticipation and experience of monetary loss. Recent theories of amygdala function argue that the amygdala subserves an abstract function in detecting uncertainty or ambiguity in the environment and in triggering arousal and vigilance as a consequence. For instance, monkeys whose amygdala has been lesioned approach stimuli that healthy monkeys avoid, and people with more inhibited personalities show greater amygdala activation than average. Losses are a possibility in many risky decisions, and organisms have evolved mechanisms to evaluate and avoid them. Research suggests that people often avoid risks with losses even when they might earn a substantially larger gain, a behavioral preference termed loss aversion. Across many studies, losses typically loom about twice as large as gains; for instance, people will often avoid gambles in which they are equally likely to either lose $10 or win $15, even though the expected value of this gamble is positive at $2.50, while they are more likely to take a gamble in which they have equal probability of losing $10 or gaining $20.

The two main participants in the study by De Martino and collaborators presented with rare focal bilateral lesions of the amygdala. Both individuals had impairments in processing fear despite otherwise largely normal cognition and IQ. The first participant, S.M., was a forty-three-year-old woman with a high school education whose lesions encompassed the entire amygdala plus subjacent white matter and part of the anterior cortex. The second participant, A.P., was a twenty-three-year-old woman with a college education whose lesions were entirely confined to the amygdala, encompassing roughly 50 percent of the amygdala. The two participants were each compared to a separate group of six healthy controls (for a total of twelve controls) matched to that participant on age, gender, monetary income, and education. The participants and controls were asked to accept or reject a series of mixed gambles with equal (or 50 percent) probability of winning or losing a variable amount of money for a total of 256 gambles. These gambles were presented on a computer screen as the prospective outcomes of a coin flip. Participants indicated their willingness to take the gamble by pressing a key. Both amygdala-lesioned participants showed a dramatic reduction of loss aversion, yet they retained a normal response to reward magnitude. This pattern of behavior is consistent with evidence that monkeys with amygdala lesions maintain a stable pattern of preference among sets of food items, even though they will approach foods that are paired with potentially threatening stimuli more quickly than control monkeys.

Further analysis showed that the lesion participants did not have an increased appetite for risk per se, because they disliked increased outcome variance and risk as much as the matched controls did. Instead, they differed only in their higher willingness to accept mixed gain-loss gambles, which is evidence of a specific reduction in aversion to loss. It was notable that whereas participant A.P. was essentially loss neutral, participant S.M. showed a mild loss-seeking behavior. This difference between the two lesion participants was mirrored by differences in their respective matched control groups. Differences in several sociodemographic factors may partly account for this variation, but the fact that S.M. has more extensive amygdala damage than A.P. may also, in part, account for the difference between the two lesion participants. Based on these observations, the scientists concluded that the amygdala must be playing a necessary role in generating loss aversion during human decision making.

The extraordinary variation in the morphology and function of brain structures among humans accounts not just for some of humanity's greatest achievements but also for some of its greatest ailments. It turns out that some people have amygdalae that are much more reactive than average when they are confronted with adverse life circumstances, and this reactivity is controlled by a genetic variation in the promoter region of the serotonin transporter gene (5-HTTLPR). Carriers of one or two short (s) allele at this locus, resulting in reduced transcriptional efficacy and protein expression, exhibit both increased resting-state amygdala blood flow and greater amygdala reactivity to emotional stimuli relative to carriers who are homozygous for the long (l) allele (meaning they have two copies of the l allele). Serotonin is both a neurotransmitter in the brain, where it plays an important role in the regulation of emotion, mood, cognition, and learning, and a hormone in the intestinal tract, where it helps with diverse physiological processes such as appetite, digestion, vasoconstriction, and vomiting. Depressed people have low levels of serotonin in their brains, and drugs that increase serotonin concentration in the synaptic plasma by inhibiting its excessive reuptake by transporters (called selective serotonin reuptake inhibitors or SSRIs) are the most commonly prescribed treatment option for depression.

Based on previous research findings linking variations in 5-HTTLPR with variations in individual amygdala reactivity and given the amygdala's involvement in the generation of fearful salience, Avshalom Caspi, from King's College London, and collaborators surmised that variations in 5-HTTLPR might moderate psychopathological reactions to stressful life experiences. They tested this hypothesis among members of the Dunedin Multidisciplinary Health and Development Study, which assessed a representative birth cohort of 1,037 children every two to three years from age three to age twenty-one on a number of factors. Of those individuals, 847 remained in the study by age twenty-six. The researchers divided them into three groups on the basis of their 5-HTTLPR genotype: 17 percent had two copies of the s allele (the s/s homozygotes), 51 percent had one copy of the s allele (the s/l heterozygotes), and 31 percent had two copies of the l allele (l/l homozygotes). There was no difference in genotype frequencies between the sexes. Stressful life events occurring after the twenty-first birthday and before the twenty-sixth birthday were assessed with the aid of a life-history calendar to screen for fourteen different adverse life events that included stressors related to employment, financial changes, housing, health, and interpersonal relationships: 30 percent of the study members experienced no stressful life events; 25 percent experienced one event; 20 percent, two events; 11 percent, three events; and 15 percent, four or more events. There were no significant differences between the three genotype groups in the number of adverse life events they experienced, suggesting that the 5-HTTLPR genotype did not influence exposure to stressful life events.

Study members were assessed for past-year depression at age twenty-six: 17 percent of study members (58 percent female versus 42 percent male) met criteria for a past-year major depressive episode, which is comparable to age and sex prevalence rates observed in US epidemiological studies. In addition, 3 percent of the study members reported past-year suicide attempts or recurrent thoughts about suicide in the context of a depressive episode. The interaction between 5-HTTLPR and life events showed that the effect of life events on self-reports of depression symptoms at age twenty-six was significantly stronger among individuals carrying an s allele than among l/l homozygotes. The analysis also showed that stressful life events predicted suicide ideation or attempt among individuals carrying an s allele but not among l/l homozygotes. The study additionally revealed that childhood maltreatment predicted adult depression only among individuals carrying an s allele but not among l/l homozygotes. Although carriers of an s 5-HTTLPR allele who experienced four or more adverse life events constituted only 10 percent of the birth cohort, they accounted for almost one-quarter of the 133 cases of diagnosed depression. Moreover, among cohort members suffering four or more stressful life events, 33 percent of individuals with an s allele became depressed, whereas only 17 percent of the l/l homozygotes developed depression. This study thus provides evidence of a gene-by-environment interaction, in which an individual's response to environmental insults is moderated by his or her genetic makeup. Individuals possessing the s allele of 5-HTTLPR appear to have highly reactive amygdalae, which increases their vulnerability to depression in the event of adverse life circumstances. Having a depressive episode makes people more likely to experience another depressive episode in the future if further life challenges are encountered, which, in turn, increases the probability of a third episode even more, in a seemingly unending cycle.

The progressively increased likelihood of experiencing a depressive episode is presumably due to alterations that happen in the brain as a result of each episode in a process called brain plasticity or neural plasticity, in which the brain structure is altered in a permanent or semipermanent way via environmental influences. In this instance, the amygdalae are altered in a way that makes them ever more sensitized and reactive to negative life experiences. Although the short 5-HTTLPR variant is too prevalent in the general population for discriminatory screening (given that over half of the Caucasian population has an s allele), a microarray of genes might eventually identify those needing prophylaxis against life's stressful events. Considering that depression is among the top five leading causes of disability and disease burden throughout the world, timely interventions guided by genetic predisposition to mitigate environmental risk for vulnerable individuals might eliminate much unhappiness globally.

Another prominent example of a gene-by-environment interaction that leads to long-lasting neural changes in the brain, or neural plasticity, is the discovery in the 1980s that the brain's dopamine systems can be enduringly sensitized by many drugs of abuse (cocaine, amphetamine, heroin, alcohol, nicotine, etc.), not just stimulated while those drugs are actually in the body. Sensitization is especially likely if the drugs are taken repeatedly and at high doses spaced apart. Neural plasticity was once thought by neuroscientists to only occur in childhood, but research in the late twentieth century showed that many aspects of the brain can be altered (or are "plastic") even throughout adulthood, although a developing child's brain exhibits a higher degree of plasticity than an adult's brain.

According to a 2016 article by Kent Berridge and Terry Robinson, from the University of Michigan, further research into addiction has revealed that drug sensitization alters the morphology of glutamate neurons that project from the cortex to the nucleus accumbens, permanently modifying their interactions with dopaminergic neurons. Furthermore, drug sensitization changes the physical structure of the neurons in the nucleus accumbens, such as altering the shape and number of tiny spines on their dendrites, which act as their "receiving antennae" for incoming signals. Functionally, this sensitization renders the brain's desire systems hyperreactive to drug cues and contexts, thus conferring more intense incentive salience on those cues or contexts. Consequently, addicts have stronger cue-triggered urges and intensely want to take drugs. Sensitized wanting can persist for years, even if the person cognitively does not want to take drugs and does not expect the drugs to be very pleasant, and even long after withdrawal symptoms have subsided. This is the reason why some individuals have near-compulsive levels of urge to take drugs and remain vulnerable to a persisting risk of relapse even after a significant period of drug abstinence. The sensitized brain produces pulses of heightened dopamine release, brain activations, and motivation that last seconds or minutes when drug contexts are encountered. This means that surges of intense wanting are most likely to be triggered when drug cues are encountered (or imagined) in contexts previously associated with taking drugs, such as viewing photos of drug paraphernalia or seeing other people taking drugs.

Compensatory neural suppressions (for instance, receptor down-regulation) do accompany heavy drug use while drug-taking continues. Suppressions produce tolerance to drug highs (and to the aversive effects of some addictive drugs—which permits the person to take higher doses, inducing even more tolerance). Neural suppressions also produce withdrawal for a while, once the drug is finally stopped. Suppressions are partial compensatory responses of the brain to the high levels of nucleus accumbens stimulation induced by drugs, essentially a temporary cellular effort by neurons to turn down their levels of chemical overstimulation. Many tolerance-related suppressions are apt to fade within weeks if drug taking is stopped. By contrast, the neural changes that cause incentive sensitization do not fade over months of drug abstinence—if anything, sensitization grows for some time during abstinence, a phenomenon sometimes called incubation of drug craving, which is an increase in relapse vulnerability after a month or so of drug abstinence.

It turns out that individuals differ substantially in their vulnerability to addiction, and most people who take drugs never become addicts. For example, only about 30 percent of people who use cocaine actually go on to become long-term addicts. Accordingly, individuals also differ considerably in their susceptibility to nucleus accumbens sensitization, even when exposed to the same drugs and doses. Genetic factors are important determinants of susceptibility to sensitization in rodents, and genes also contribute strongly to addiction vulnerability in humans. Other determinants of sensitization vulnerability include gender, the presence of sex hormones, and whether the individual has had major stresses in life before taking drugs. Individuals with combinations of these several factors may be most at risk of developing incentive sensitization to drugs of abuse and addiction. Finally, among those who experiment with drugs in the first place, important situational factors can facilitate incentive sensitization or, alternatively, make it less likely. These include how long the drugs have been used, whether the dosage has been escalated, and whether the person took the drugs by routes that resulted in the drugs rapidly reaching the brain, as is the case in intravenous use compared to inhalation.

Various behavioral addictions have gained attention in recent years that do not involve drugs at all: eating addiction, gambling addiction, sex or pornography addiction, internet addiction, shopping addiction, and so on. Evidence is emerging that individuals with these behavioral addictions may have some sensitization-like patterns of brain hyperreactivity to cues related to their own personal addictions. Proof for the idea that diverse compulsive motivations such as gambling can be caused by overstimulation of dopamine-related systems comes from studying Parkinson's patients, about 15 percent of whom develop what is called dopamine dysregulation syndrome when treated with newer dopamine-stimulating medications that directly activate their brain dopamine receptors. These patients can develop intense compulsive motivations to pursue gambling, shopping, sex, internet use, eccentric hobbies, or similar activities. Some patients may even take excessive amounts of their medication in a more classical drug-addiction fashion. Usually, the compulsive motivations rapidly fade if the dopamine-stimulating medications are stopped. In treating drug addictions, psychological approaches, such as cognitive and behavioral therapies, twelve-step programs, contingency management, and mindfulness therapy, arguably remain more effective than any medications available today. Recent findings from animal studies give some reason to hope that effective sensitization-reversing treatments might be developed in the future.

Research continues to reveal how normal incentive and fearful salience processes in the brain's desire-and-dread circuitry can become distorted, how their distortion can trigger psychological disorders and affect personal happiness, and how new therapies might someday overcome distortions to improve function again. Rewards can be both liked and wanted. The brain circuitry that mediates the psychological process of wanting (or desiring) a particular reward not currently possessed is dissociable from the circuitry that mediates the degree to which it is liked (or pleasurable) once it is obtained. The same is true of punishments. The brain's relatively large and robust dopamine system is in charge of assigning incentive salience to rewards and the appetitive motivation to approach them, and fearful salience to punishments and the aversive motivation to avoid them. By comparison, the pleasurable impact of the rewards that we have obtained and the pain of the punishments inflicted upon us in the here and now are mediated by smaller and more fragile neural systems, and neither pleasure nor pain are dependent on dopamine. Perhaps it is this difference in the relative size of the brain's desire-and-dread circuitry compared to its pleasure-and-pain circuitry that, at least in part, explains why anticipating a reward can be a much more powerful experience than actually consuming it and why the anticipation of a punishment can feel so much worse than experiencing it.

3

Pleasure and Pain

Two opposing philosophical traditions have suggested ways to reach happiness. Hedonism advocates that happiness is achieved through pleasure, enjoyment, and the avoidance of pain, while eudaemonism contends that happiness is only reached through complex and meaningful goals, the achievement of which may involve putting oneself through occasional hardship and pain. The word *hedonism* is derived from a Greek term meaning pleasure, while *eudaemonism* is from a Greek word translating to the state or condition of good spirit or well-being. Democritus seems to be the earliest philosopher on record to have categorically embraced a hedonistic philosophy. He contended that cheerfulness and contentment were the supreme goals of life and further claimed that joy and sorrow are the distinguishing marks of things beneficial and harmful. Epicurus later followed in the steps of Democritus and advocated for a version of hedonism according to the Golden Rule; he believed that the greatest good was to seek modest, sustainable pleasure in the form of a state of tranquility and freedom from fear and absence of bodily pain through knowledge of the workings of the world and the limits of our desires. Eudaemonia, on the other hand, is a central concept in Aristotelian ethics.

In *The Nicomachean Ethics*, Aristotle posited that a eudaemonic life is one of "virtuous activity in accordance with reason." It is important to bear in mind that the sense of the word *virtue* operative in ancient ethics is not exclusively moral and includes more than states such as wisdom, courage, and compassion. It also incorporates the notion of excellence. In this sense, speed is a virtue in a racehorse, height and physical prowess are virtues in a basketball player, and mastery of geometry is a virtue in a mathematician. In practice, a state of eudaemonia results from striving to achieve complex and meaningful life goals, resulting in both gradual improvements of the self and significant contributions to society. While hedonism and eudaemonism have been construed as opposing philosophies at times, the modern definition of happiness includes both. The most common definition of happiness today has a hedonistic or affective component (the idea of maximizing positive affect and minimizing negative affect in one's life), as well as a eudaemonic or cognitive component (in the form having a positive rating of one's overall satisfaction with life, which, to a great extent, results from the accomplishment of personally important goals).

The distinction between eudaemonism and hedonism can also be interpreted as the reflection of the two separate brain systems that humans have evolved to process things in the extrapersonal space (meaning things that are distant in time and space), which we called the desire-and-dread circuitry of the brain, versus things that are in the peripersonal space (meaning things that are in the here and now), which we are calling the pleasure-and-pain circuitry of the brain. The former processes the anticipation of future pleasure or pain, resulting in the thrill of a delayed reward or the dread of a delayed punishment, while the latter processes the current experience of pleasure and pain, resulting in the enjoyment or liking of a current reward or the suffering inflicted by a presently experienced punishment or adverse event.

In the 1980s and before, most scientists generally accepted the idea that dopamine mediates reward pleasure, meaning the hedonic impact of tasty food, addictive drugs, sex, and other rewards. Many studies had found that brain dopamine systems were activated by most rewards and, further, that manipulating dopamine altered wanting for rewards—for example, changing how much laboratory animals preferred, pursued, worked for, or consumed the reward. Changes in wanting were naturally interpreted to reflect corresponding changes in liking, based on the assumption that wanting was proportional to liking. This belief led Kent Berridge, Isabel Venier, and Terry Robinson to do an experiment on rats in 1989, in which they hypothesized that depletion of brain dopamine via a neurochemical lesion in rats would reduce liking reactions for pleasant tastes. They used a naturalistic assessment of sweetness-induced pleasure in the animals based on affective facial expressions of liking. Sweetness typically elicits a relaxed countenance and rhythmic tongue and mouth expressions of liking, whereas bitterness elicits disgust, gapes, and turning away. These affective facial expressions to taste are homologous in human infants, apes, monkeys, and even rats. Berridge and collaborators were thus expecting that rats whose brain dopamine had been depleted via lesions would display reduced facial expressions of liking in response to food rewards. But that is not what they found. Instead, they were surprised to find that liking reactions of rats to sugar taste were completely normal even after depletion of nearly all brain dopamine. The dopamine lesions did apparently abolish all motivation—the rats displayed profound aphagia, as they no longer sought or worked to consume food rewards. When the sugary liquid was placed directly in their mouths, however, the rats clearly exhibited facial expressions indicative of liking. To make sense of these paradoxical findings, they proposed that the brain's dopamine systems mediate wanting or desire (in particular, a psychological process called incentive salience), but not liking for the same reward. Otherwise stated, the capacity for hedonism can be neurologically dissociated from motivated appetitive behavior. A follow-up study using implanted electrodes to stimulate the same systems and raise dopamine levels also failed to enhance facial expressions of liking, despite quadrupling a rat's wanting (or willingness to work) to eat food rewards. This subsequently led to the abandonment of the idea that dopamine was the "pleasure molecule," leaving scientists in a quandary. If not dopamine and its associated brain circuitry, what then mediates pleasure?

Insight came from additional research. According to a 2009 publication by Kent Berridge, affect begins in the brain stem. The brain stem is the oldest and innermost region of the brain. It connects the cerebellum to the spinal cord and handles all communication between our body and brain. It has crucial integrative functions, being involved in cardiovascular system control, respiratory control, pain sensitivity control, alertness, awareness, and consciousness. Thus, brain stem damage is a very serious and, often, life-threatening problem. Basic brain stem circuits participate in liking reactions as well as in pain and are partially autonomous, able to function as reflexes in the brain stem alone. For example, basic positive or negative facial expressions are still found in human anencephalic infants, who are born with a midbrain and hindbrain but no cortex, amygdala, or classic limbic system, due to a congenital defect that prevents prenatal development of their forebrain. Yet, sweet tastes still elicit normal positive affective facial expressions from anencephalic human infants, whereas bitter or sour tastes elicit negative expressions. Similarly, a decerebrate rat has an isolated brain stem because of a surgical transaction at the top of its midbrain that separates the brain stem from the forebrain, but that isolated brain stem remains able to generate normal positive expressions to sweet tastes and negative expressions to bitter tastes when those are placed in the rat's mouth.

Certainly, a decerebrate rat or an anencephalic infant cannot like a sweet taste in the same sense that a normal individual does. They may not experience the cognitive subcomponent of liking, but they should have a capacity to feel a residual affective subcomponent of it. After all, almost every feeling of physical pleasure or pain felt by our forebrain has climbed its way there through the brain stem. Ascending signals do not just pass through the brain stem; much processing happens to them on the way up. There is compelling reason to believe that affect begins in the brain stem for both pleasure and pain. In a normal brain, brain stem sites make important contributions to affective experiences that are, for the most part, generated by forebrain circuits just above them. For instance, even decerebrate rats show enhanced liking reactions to sucrose taste after benzodiazepine administration directly into the parabrachial nucleus of the pons, a component of the brain stem and one of a number of small hedonic hotspots in the brain. Benzodiazepines are a type of medication known as tranquilizers. Familiar commercial names include Valium and Xanax. They are some of the most commonly prescribed medications for the treatment of anxiety and insomnia in the United States. Even a decerebrate brain may contain the kernel of a liking reaction that the word reflex does not adequately capture, just as the brain stem also contains substantial circuitry for pain and analgesia. This may reflect the adaptive functions of affective reactions throughout brain evolution and may also be relevant to how unconscious liking reactions occur in people even today.

In a standard human brain, the brain stem participates more fully in liking and wanting when it is connected to the forebrain and becomes a hierarchical intermediary stage in larger affective circuits. The human brain evolved in a reiterative fashion, where a newly formed forebrain atop the brain stem and midbrain re-represented the pleasures and pains that the brain stem had already represented in a simpler fashion, and then the even more recently formed prefrontal cortex re-represented them one more time, forming continuously more complex combinations. The forebrain systems normally control the brain stem circuits, and the prefrontal cortex the forebrain systems, so that normal pleasure and pain reactions are not merely brain stem reflexes in a whole-brained individual.

Research done at the Berridge laboratory, at the University of Michigan, to pinpoint affect generating circuits in the forebrain has identified a hedonic hotspot in the nucleus accumbens that uses opioid and endocannabinoid signals to amplify liking for sweetness. Opioid neurotransmitters, such as enkephalin and endorphin, which are released by the body to alleviate pain during the fight-or-flight or stress response, are mimicked by opiate drugs made from the opium poppy, such as heroin, synthetic opioids such as fentanyl, and pain relievers available legally by prescription, such as oxycodone, hydrocodone, codeine, morphine, and many others. Endo-cannabinoid neurotransmitters—for instance, anandamide—are mimicked by cannabinoid drugs such as marijuana. Research suggests that endogenous opioid or cannabinoid receptor activation stimulates appetite in part by enhancing liking for the perceived palatability of food. This may be the reason why people will order pizza after smoking marijuana. The hedonic hotspot for opioid enhancement of sensory pleasure located in the nucleus accumbens is specifically within the rostrodorsal quadrant of its medial shell. It is about a cubic millimeter in volume in rats, and probably about a cubic centimeter in volume in humans. When researchers inject a drug that works by activating the *mu* type of opioid receptors into this hedonic hotspot in rats, the animals display more than double the usual number of positive liking reactions to sucrose taste. The same hotspot has a type of cannabinoid receptors called CB1 receptors, which are sensitive to anandamide. Microinjection of anandamide into this hotspot in rats potently doubles the number of positive liking facial reactions that sucrose taste elicits from the animals.

The same hotspot microinjections of opioids or cannabinoids also stimulate wanting or the motivation to eat food. But wanting mechanisms extend far beyond hedonic hotspots in the nucleus accumbens. For example, the opioid hedonic hotspot comprises a mere 10 percent of the entire nucleus accumbens, and even only 30 percent of its medial shell. Yet opioid microinjections throughout the entire 100 percent of the medial shell potently increase wanting, more than doubling the amount of food intake. Outside of the hedonic hotspot, opioid stimulations produce very different effects. For example, in the nucleus accumbens, at virtually all other locations, opioid microinjections still stimulate wanting for food as much as in the hotspot, but do not enhance liking (and even suppress liking in a more posterior coldspot in the medial shell, while still stimulating food intake). Thus, opioid sites responsible for liking are anatomically dissociable from those that influence wanting.

The ventral pallidum, a structure located in the forebrain in the proximity of the nucleus accumbens, is a chief target for nucleus accumbens outputs, and its posterior half contains another opioid hotspot. What seems to be an astounding fact is that the ventral pallidum and its environs contain the only brain region known so far where the death of neurons abolishes all liking and replaces it with disliking, even for sweetness (at least for several weeks). Following the death of most neurons in the hedonic hotspot of the ventral pallidum, sugar tastes bitter rather than sweet. Lesion of the ventral pallidum in rats induces extreme food aversion; the animals will starve themselves to death unless given intensive nursing care and artificial intra-gastric feeding. These lesions seem to disinhibit other aversion-generating systems of the forebrain, so what remains is disliking for everything. The ventral pallidum can also generate enhancement of natural pleasure when it is intact, by opioid stimulation of its own hedonic hotspot. When an opioid is injected in this hotspot in rats, sucrose taste elicits over twice as many liking reactions as it normally does and also causes the rats to eat over twice as much food. The activity of neurons in the posterior hedonic hotspot of the ventral pallidum appears to specifically code liking for sweet, salty, and other food rewards. In rats, recording electrodes can be permanently implanted in the ventral pallidum, and neurons there fire faster when rats eat a sweet taste. The firing of sucrose-triggered neurons appears to reflect hedonic liking for the taste. For example, the same neurons will not fire to an intensely salty solution that is unpleasant, such as a solution that is three times saltier than seawater. The neurons suddenly begin to fire, however, to the triple-seawater taste if a physiological state of "salt appetite" is induced in the rats, by administering hormones that cause the body to need more salt, and this increases the perceived liking for intensely salty taste. Thus, neurons in the ventral pallidum code taste pleasure in a way that is sensitive to the physiological need of the moment.

Scientists think that there may be more hedonic hotspots in the brain besides those identified so far in the brain stem's parabrachial nucleus of the pons and in the forebrain's nucleus accumbens shell and ventral pallidum. The hedonic hotspots distributed across the brain may be functionally linked together into an integrated hierarchical circuit that combines multiple forebrain and brain stem spots, akin to multiple islands of an archipelago that trade goods together. Otherwise stated, in a standard human brain, these hotspots are not isolated islands that function independently from one another, but they are functionally tied together into a single integrated circuit. There is, indeed, some evidence that the enhancement of liking by hotspots in the nucleus accumbens and the ventral pallidum may act together as a single cooperative hierarchy, needing unanimous "votes" by both hotspots. For example, hedonic amplification by opioid stimulation of one hotspot can be disrupted by opioid receptor blockade at the other hotspot. The orbitofrontal cortex, the part of the frontal cortex located just above the orbits of our eyes, is particularly activated in response to various pleasures, such as food, sexual orgasms, drugs, chocolate, and music. This area of our prefrontal cortex may modulate the raw pleasure signals coming from the hedonic hotspots located deeper in the forebrain and brain stem to generate our conscious experience of pleasure. A malfunction of the hedonic mechanisms in the orbitofrontal cortex could contribute to the profound changes in eating habits (such as escalating desire for sweet food and reduced satiety) that are often followed by enormous weight gain in patients with frontotemporal dementia. Thus, a single hedonic circuit may combine multiple brain locations and neurotransmitters to potentiate liking reactions and pleasure.

It is entirely plausible that the brain's pleasure circuitry may be involved in eating disorders. Indeed, one way that eating disorders can be explained is through abnormal functioning of the brain's reward systems. Foods might become liked too much or too little via reward dysfunction. For example, pathological overactivation of the opioid or endocannabinoid hedonic hotspots in the nucleus accumbens and ventral pallidum might cause an enhanced liking reaction to taste pleasure in some individuals. An endogenously produced increase in opioid tone there could in principle magnify the hedonic impact of foods, making an individual like food more than other people do and make them want to eat more, leading to obesity. Conversely, a suppressive form of hedonic hotspot dysfunction might reduce liking, or even create disliking for normally palatable food, causing anorexia nervosa.

While some individuals may be capable of deriving more pleasure from certain rewards than others, people also seem to differ dramatically in their sensitivity to pain. Both traits can affect personal happiness in profound ways. According to a 2019 article by Ian Sample in *The Guardian*, Jo Cameron, a retired teacher from Inverness, Scotland, has experienced broken limbs, cuts and burns, childbirth, and numerous surgical operations over the years with little or no need for pain relief. But it is not only an inability to sense pain that makes Jo Cameron stand out; she also never panics. When a van driver ran her off the road a few years ago, she climbed out of her car, which was on its roof in a ditch, and went to comfort the shaking young driver who cut her off. She only noticed her bruises later. She is relentlessly upbeat, and in stress and depression tests she scored zero. "I knew that I was happy-go-lucky, but it didn't dawn on me that I was different," she says. "I thought it was just me. I didn't know anything strange was going on until I was sixty-five."

The moment of realization came when Cameron had X-rays for a bad hip. Now and again her hip would give way, making her walk lopsided. For three or four years, her general practitioner, and then the hospital, turned her away because she was not in pain. When she was finally scanned, the X-rays revealed massive deterioration of the joint. "I'd not had a twinge. They couldn't believe it." Following the hip replacement, doctors noticed that her thumbs were deformed by osteoarthritis. They immediately booked her in for a double hand operation, a procedure described as "excruciating" by one surgeon. Again, Cameron felt almost no pain after the operation. A consultant, Devjit Srivastava, who was overseeing her care at Raigmore hospital in Inverness, was so stunned that he referred her to pain specialists at the University College London.

It turns out that Jo Cameron owes her bliss to two genetic mutations. The first mutation is rather common and dampens the activity of a gene called FAAH, which makes an enzyme that breaks down anandamide, a chemical in the body that is central to pain sensation, mood, and memory. Anandamide works in a similar way to the active ingredients of cannabis. The less it is broken down, the more its analgesic and other effects are felt. The second mutation was previously unknown and consisted in the deletion in a nearby gene, which the scientists named FAAH-OUT. It is thought to work like a volume control on FAAH. If disabled, FAAH falls silent. The upshot is that anandamide, a natural cannabinoid, builds up in the system. Anandamide is found in nearly all tissues in a wide range of animals, and also in plants, including in small amounts in chocolate. The word *anandamide* is from the Sanskrit word *ananda,* which means joy, bliss, or delight. The acute beneficial effects of exercise (known as runner's high) seem to be mediated by anandamide in mice. It was shown to inhibit the proliferation of certain human breast cancer cell lines in vitro and to impair working memory in rats. Anandamide also plays a role in the interpretation of stimuli; specifically, it makes one interpret an ambiguous cue more optimistically rather than pessimistically. As an analgesic, it provides pain relief. It may also decrease anxiety and induce mild euphoria. Anandamide injected directly into the nucleus accumbens hedonic hotspot enhances the pleasurable responses of rats to a rewarding sucrose taste and enhances food intake as well. Increasing anandamide seems to increase the intrinsic value of food, not necessarily stimulating appetite or hunger. The "happy-go-lucky" Jo Cameron has twice as much anandamide as those in the general population. "I was quite amused when I found out," she said. "And then they told me about these other things, the happiness and the forgetfulness. I'm always forgetting things; I always have done. It's good in lots of ways but not in others. I don't get the alarm system everyone else gets." The discovery has boosted hopes of new treatments for chronic pain, which affects millions of people globally. Who knew that personal happiness can sometimes be the result of a couple of lucky genetic mutations?

As fate would have it, those of us who feel less physical pain also suffer less socially, and vice versa. According to research done by Matthew Lieberman and Naomi Eisenberger, from the University of California at Los Angeles, the brain uses the same circuits to handle the social and physical variants of pain and pleasure. While desire and dread emanate from the same brain circuitry that tags stimuli that are distant in time and space as appetitive or aversive, pleasure and pain seem to stem from separate circuits when it comes to experiencing rewards and punishments in the here and now. Acute physical pain serves as a protective mechanism that alerts us to potential tissue damage and drives a behavioral response that removes us from danger. The neural circuitry critical for mounting this behavioral response is situated within the brain stem, the most ancient and deepest part of our brain.

Four brain stem structures are particularly involved in the processing of physical pain: the periaqueductal gray, rostral ventromedial medulla, locus coeruleus, and subnucleus reticularis dorsalis. The pain signals from these brain stem structures are further processed in the forebrain by the anterior part of the ventral pallidum, and then relayed through the thalamus to the cortex for further processing to produce our conscious experience of pain. The brain's physical pain system involves three regions of the cortex. The somatosensory cortex codes for where in the body the pain is located, particularly for pain felt on the skin. The insula, which also mediates our disgust response, is particularly responsive to internal and visceral forms of pain and seems to provide the brain with a representation of the body's overall state. Last, the dorsolateral anterior cingulate cortex is the region that is associated with the conscious distress that we feel in response to physical pain. This region is associated with how much a particular pain stimulation bothers a person.

People who experience panic attacks have a hyperreactive locus coeruleus, which is one of the brain stem nuclei that maintains close communications with the amygdala. Stimulating an animal's locus coeruleus, just like stimulating its amygdala, produces anxiety behaviors. Substances with tranquilizing effects, such as benzodiazepines, alcohol, and opiates, make the locus coeruleus less active. People with panic and anxiety disorders are especially vulnerable to developing addictions to these substances since they often try to self-medicate to numb their psychological pain. Jo Cameron may have brain circuitry that is capable of generating pain signals comparable to the brain of an average individual. She has a very high concentration of the powerful analgesic anandamide in her brain, however, and this biochemical immediately suppresses the pain signals. Another way that a person could be insensitive to pain is if they had a relatively nonreactive pain circuitry.

Since there was evidence from research on social animals such as primates that the dorsolateral anterior cingulate cortex is also involved in social pain, Lieberman and Eisenberger wanted to know whether that was also true of humans. They convinced their experimental test subjects that while they were laying in a brain scanner in one location of the campus, two other subjects were laying in scanners elsewhere on the campus, and that the three of them were going to play a game together over the internet while their brains would be scanned. In reality, there were no other players, as participants played with a preset computer program. But telling them this fabrication was a way of making sure that participants truly experienced the game as social. What they found was striking. Participants showed greater activation of their dorsolateral anterior cingulate cortex when they were excluded compared to when they were included. Moreover, those who reported feeling more social pain in general also showed greater activity in the same cortical region during the exclusion phase of the game. Grieving over the death of a loved one and being treated unfairly by others are also known to activate this brain area.

Why would the brain use the same pain circuitry to process both physical and social pains? The researchers believe that the social pain system may have piggybacked on the physical pain system during mammalian evolution, borrowing the pain signal to indicate broken social bonds. Although people generally agree that our basic survival needs are food, water, and shelter, because the mammalian young are born in a particularly vulnerable state, incapable of providing for their own physical needs, they must stay connected to their caregivers. In young mammals, the need for this social connection usually supersedes the need for food, water, and shelter, because without a caregiver to provide for those needs, young mammals would not survive. Just as evolution has wired us to feel pain when we lack food via the feeling of hunger, water via thirst, or shelter via freezing or sunburn, perhaps evolution has wired us to feel pain in the form of anxiety, depression, and psychosomatic concerns when we lack or anticipate the lack of social connection. Research shows that the brain also uses the same reward circuitry to respond to social rewards in the form of social recognition or praise as it does to respond to one of the most ubiquitous human reward, namely money. Having a good reputation, being treated fairly, and being cooperative all activate the nucleus accumbens in the forebrain. Strikingly, making charitable donations activates the reward network more than receiving the same sum of money for oneself. Accordingly, we may need to appreciate that however much reality we accord to physical pain should also be extended to social pain. On the other side of the equation, when our physical needs are met, we feel pleasure. Eating when hungry, drinking when thirsty, and coming in from the cold are all pleasurable experiences that occur while a physical need is being satiated. The need for social regard may seem less immediate and real than these needs, but it is nonetheless a basic need. When our social needs are being satisfied, the brain responds in much the same way as it responds to more tangible rewards. Our attentiveness to the social world may sometimes seem like a diversion from more concrete concerns, but increasingly, neuroscience is revealing ways in which such attention is actually an adaptive response to some of our most vital concerns.

As universal as the need for social connectedness may be, people can differ strikingly in the strength of this need. Those who score especially high on the personality trait of agreeableness tend to feel a much stronger urgency for social connectedness and harmony than those who score lower on this personality factor. Two neurotransmitters that are among the brain's endogenous opioids, oxytocin and vasopressin, are in charge of mediating human social bonds in females and males, respectively. In humans, just as in other animal species, oxytocin and vasopressin contribute to a wide variety of social behaviors beyond mediating bonding, including social recognition, communication, parental care, and territorial aggression.

For instance, voles are a type of bird with two distinct populations: prairie voles and montane voles. Prairie voles are socially monogamous; males and females form long-term pair bonds, establish a nest site, and rear their offspring together. In contrast, montane voles do not form a bond with a mate, and only the females take part in rearing the young. Research shows that the monogamous behavior of the prairie voles may be due to the high levels of oxytocin receptors in the hedonic hotspot of their nucleus accumbens and high numbers of vasopressin receptors in the hedonic hotspot of their ventral pallidum. These receptor concentrations are controlled by the length of a gene called AVPR1A. The long version of this gene leads to high concentrations of oxytocin and vasopressin receptors and monogamous behavior, while the short version results in low concentration of these receptors and promiscuous mating behavior. Similar genetic variation in the human AVPR1A gene may contribute to variations in human sexual and social behavior. For instance, extreme deficit of oxytocin and vasopressin receptors in the brains of people with severe autism disorder can result in social withdrawal, lack of empathy, and an innate inability to bond with others. This can have devastating consequences for the emotional well-being of the affected individuals and their families. Autism disorders come on a spectrum with varying levels of severity. A high-functioning autistic individual may have little interest and ability in forming affiliative relationships, but they can usually form agentic relationships, which involves relating to others in order to accomplish a specific goal. While people with autism spectrum disorders may score especially low on the agreeableness trait, individuals with a diagnosis of adult separation anxiety disorder tend to score especially high on this personality factor. These individuals have extreme anxiety about separations, actual or imagined, from major attachment figures. They feel overwhelmed by the idea of having to face life's challenges on their own, so they cling to another adult they perceive to be stronger than themselves. In many instances, just like it is the case for children, the attachment figure is their way of obtaining basic life's necessities, such as food, shelter, and clothing. Sometimes these clingy behaviors can also be observed in professionally accomplished individuals, and this puts them at risk for financial exploitation by others. Indeed, their fear of being alone is so strong that they will usually do anything in order to keep their dependent relationship intact, and this includes tolerating sexual, verbal, or physical abuse from their attachment figure. Thus, extreme scores on the personality trait of agreeableness represent a vulnerability to maladaptive social bonding behaviors and their associated psychopathologies.

Oxytocin plays a particularly strong role in human bonding since it is the primary agent in helping to form the bond between a mother and her infant, and its effect has been expanded to encompass all other types of human bonding, including the attachment between lovers, family members, and friends. Oxytocin is released during all human interactions aimed at bonding, albeit at different levels depending on the nature of the relationship. At the chemical level, human social attachment is very similar to opioid addiction. Indeed, oxytocin is the endogenous homologue of heroin. They resemble each other not just in their chemical structures but also in their effects on the human brain. They both reduce anxiety, induce mild euphoria, produce profound dependence, and result in painful withdrawal symptoms when the drug—whether it be endogenous or synthetic—is stopped. Thus, broken human bonds can result in pain that very much resembles what an addict feels when they stop using heroin, especially when the bonds were particularly strong. Since women tend to have much higher levels of oxytocin than men, less testosterone to suppress its effect, and a higher concentration of oxytocin receptors in their brains, it should not come as a surprise that the majority of people with adult separation anxiety disorder are female. Because males tend to lack both oxytocin receptors and oxytocin itself (due to a combination of low production levels and the silencing of its effect by testosterone), they are more than twice as likely as females to present with autism spectrum disorders. In time, therapies consisting of either enhancing or suppressing the amount of oxytocin in the brain or the brain's sensitivity to it may be developed to help alleviate much social unhappiness currently plaguing certain individuals.

Oxytocin and vasopressin are very similar neurotransmitters chemically, having evolved from a common root. They have specialized over time to regulate the differential reproductive bonding roles that women and men have traditionally played: nursing newborns for females versus providing food and shelter for the family for males. Each is produced by both sexes, albeit in different quantities. While the excess of the sex hormone testosterone works to inhibit the action of oxytocin and enhance that of vasopressin to a certain extent in men, the excess of the sex hormone estrogen has the opposite effect in women. Both neurotransmitters not only have analgesic physical effects, but they also present with psychologic effects to up-regulate well-being and down-regulate stress and anxiety. In humans as in other animals, oxytocin and vasopressin work to promote healing and effective social functioning, in addition to inducing positive affect. They trigger an emotional sense of safety and high levels of social sensitivity, as well as an urge to be prosocial, meaning a desire to engage in behaviors that benefit other people or society as a whole, such as sharing, donating, cooperating, and volunteering.

Everybody has had an experience in which slow-tempo music and a pleasant fragrance in a warm room made us relax. Most of us are familiar with the feeling of well-being after yoga, a massage, or a meditation session. It is the release of opioid neurotransmitters in the brain that contributes to this sense of relaxation, trust, psychological stability, and reduction of stress responses, including anxiety, that we enjoy in these situations. Without these neurotransmitters and without the ability to form attachments, the human brain as we know it could not exist. Selective relationships and attachments are central to human health and well-being, both in current societies and during the course of evolution. The presence or absence of social bonds has consequences across the lifespan. Researchers are looking into the possibility of retuning the nervous system in people such as the severely autistic, whose brains lack these feel-good biochemicals or are insensitive to them, through early exposure to social experiences, as well as to oxytocin and vasopressin, hoping to alter their biologically set thresholds for sociality, emotion regulation, and aggression at developmentally critical stages. Alternatively, oxytocin receptor blockers might help those women who will tolerate severe abuse at the hands of others because of excessive production of oxytocin in their brains or excessive sensitivity to it.

The people who are near and dear to us can make us either happy or miserable, depending on the nature of the relationship. For instance, receiving negative social evaluation, such as a negative job performance feedback, particularly by a superior, can conjure negative emotions and elicit social anxiety. Indeed, negative social feedback can affect mood, self-esteem, behavior, physiology, and even lead to negative effects on both mental and physical health in case of repeated or prolonged experience. Furthermore, the impact and power of negative feedback are often more potent than positive feedback, especially in those who score high on the personality factor of neuroticism, as they are naturally more sensitive to threats and punishments than a person with an average neuroticism score. Clinically, fear of negative evaluation is a prominent feature of social anxiety displayed in social anxiety disorder, which may be comorbid with other disorders, such as autism spectrum disorder and depression. It has also been associated with delusional ideation in schizophrenia.

There is research pointing at the role of oxytocin and vasopressin in modulating the reactivity of both the amygdalae and the prefrontal cortex to negative social and emotional stimuli. These research findings motivated Marta Gozzi, from the National Institute of Mental Health in Bethesda, Maryland, and her collaborators to conduct an experiment in which twenty-one healthy men underwent functional magnetic resonance imaging in a double-blind, placebo-controlled, crossover design to determine how intranasally administered oxytocin and vasopressin modulated neural activity when receiving negative feedback on task performance from a study investigator. They found that under placebo (which is usually a sugar pill), a preferential response to negative social feedback compared with positive social feedback was evoked in brain regions putatively involved in theory of mind, pain processing, and identification of emotionally important visual cues in social perception. Interestingly, these activations weakened with oxytocin and vasopressin administration such that neural responses to receiving negative social feedback were not significantly greater than positive social feedback. These findings are important because they show that exogenous administration of oxytocin and vasopressin could be used to help those who have disorders with impairments in social information processing, such as those suffering from severe autism, depression, social anxiety, or schizophrenia.

While the brain levels of the neurotransmitters that allow us to feel pleasure and pain in response to immediate rewards and punishments and the biological design of the receptors and transporters that are sensitive to them matter a great deal to our well-being or ill-being, the way that our prefrontal cortex interprets the signals from the deeper and more ancient brain structures—the brain stem and the limbic brain—and what it allows or disallows into consciousness ultimately determine the quality of our subjective experience of reality. Research shows that what we perceive is not necessarily what there is, and two people can experience the very same thing in utterly different ways. Our happiness or unhappiness are critically dependent upon the nature and strength of the cognitive control that our prefrontal cortex is capable of exerting upon the deeper brain structures and its ability to learn and readjust its responses to environmental situations.

4

Cognitive Control and Learning

According to Daniel Gilbert, human beings come into this world with a passion for control, and they go out of the world the same way, and research suggests that if they lose their ability to control things at any point between their entrance and their exit, they become unhappy, helpless, and depressed. The ability to control our experience is the domain of the prefrontal cortex, the latest addition to the human brain, and it is this brain structure that distinguishes us most from the lower primates. It is the part of the brain that develops the latest in youth and decays the first in old age. It performs functions that are critical to consciousness, such as directing attention, chunking (which means grouping together connected items or words so that they can be processed or stored as single concepts), keeping representations and concepts in working memory for manipulation in decision making, and goal-directed action. Humans have evolved consciousness for the purpose of providing solutions to otherwise intractable problems. Consciousness involves attending to information relevant to a biological goal, especially when automatic unconscious routines are unable to carry out that goal. Automatic or instinctive responses have been etched into our brains by the forces of evolution and are appropriate whenever the perception of a given stimulus or situation always calls for the same action. No analysis is required. What matters is the quick matching of an environmental condition to a given response. More complex situations or tasks, where multiple features need to be managed simultaneously or in series via working memory, are those that require consciousness. The prefrontal cortex is in charge of detecting and encoding regularities between items in working memory to help understand a complex situation, so that a given goal is more likely to be reached.

The conscious experience of a sensory event, such as how hot or cold an object can feel to our hand, is derived from a complex integration that takes place in the prefrontal cortex and includes not only information arising from peripheral sensory transducers but also cognitive information about the present context, past history, and future implications of what we are sensing. This makes the sensory experience of a given stimulus unique to each individual, a fact that is especially pertinent to the experience of pain. For instance, some people claim that they are very sensitive to pain, whereas others say that they tolerate pain well. Yet, it is difficult to determine whether such subjective reports reflect true experiential differences between individuals. Undetectable physical differences in injuries or disease processes can result in chronic pain for one individual but only minimal deficits for another. Furthermore, an individual's subjective experience of pain can vary substantially from day to day despite being evoked by a constant stimulus. Does the person's self-reported pain simply vary with their momentary emotional state, unrelated external events such as the presence or absence of sunshine, or is it correlated to a physical phenomenon actually happening in their body that is measurable?

Objectively assessing how much people can differ in their ratings of pain from the same noxious stimulation is especially important for physicians involved in pain management who must rely on patients' self-reports for treatment. Equally important is knowing whether the variability in the reported levels of pain emanate from differential sensitivity of spinal and peripheral sensory transducers or from differential cognitive processing of these lower order signals. To gain insight into these topics, Robert Coghill, from Wake Forest University, and his collaborators recruited seventeen subjects (eight women and nine men) between the ages of twenty-one and forty who gave consent to undergo brain imaging and experience experimental pain stimuli. The pain stimuli were first delivered in the form of noxious heat using a thermal device set to 120 degrees Fahrenheit and applied to their nondominant ventral forearm for a set duration. Subjects provided a description of the time course of pain intensity by continuously rating these stimuli, which were identical to those presented during the functional imaging session of their brains that took place later. Each subject's responses were normalized to a range of zero to ten and averaged together to characterize the time course of perceived pain intensity. Post-stimulus ratings of pain intensity were also acquired for these stimuli to enable direct comparison with those obtained during the scanning session, which followed next. Each of the subjects underwent functional brain imaging during thermal stimulation of the skin of their right lower leg. The device was set to 120 degrees Fahrenheit for the same duration as during the previous session. At the end of the functional imaging series, subjects provided a psychophysical rating of pain intensity.

Interestingly, the subjects' ratings of the intensity of pain felt on their right lower leg differed substantially despite the fact that the very same thermal device, set exactly at the same temperature and duration of delivery, was used on all of them. The most sensitive subject rated the noxious heat as 8.9 out of ten, whereas the least sensitive individual rated the same stimulus as 1.05 out of ten! There was a strong match between the ratings of each individual in the session where the heat was applied to their forearm compared to the scanning session, which took place several days later and where the heat was applied to their right lower leg. The ratings from the scanning sessions were used to assign subjects to a high-, moderate-, or low-sensitivity subgroup. Next, the researchers compared the brain scans of people in the different subgroups. Within each subgroup, they saw similar levels of activation of three cortical areas involved in processing pain: the primary somatosensory cortex (involved in localizing the pain), the anterior cingulate cortex (involved in assigning the negative emotional valence to the experience of pain), and the prefrontal cortex (involved in setting the perceived pain intensity). All three of these cortical regions exhibited significantly greater magnitudes of activation in the high-sensitivity subgroup than the low-sensitivity subgroup.

The thalamus is a brain structure that is in charge of relaying information from the spinal cord and brain stem to the cortex for further processing. The brain scans showed that neither the frequency nor magnitude of thalamic activation was significantly different between the high- and low-sensitivity subgroups. These two results taken together provide a demonstration that the three cortical regions that were activated in response to painful stimulation may be critically important in processes leading to between-individual differences in pain sensitivity. Given that the thalamus is the primary relay for afferent transmission of nociceptive information, the absence of detectable brain imaging differences in the thalamus (in combination with robust differences in the aforementioned three cortical regions) suggests that generally similar afferent input was conveyed to thalamic levels in both high- and low-sensitivity individuals. Accordingly, a large portion of the differences between individuals in both the subjective experience of pain and activation of their cortex is likely attributable to factors other than differential sensitivity of spinal or peripheral afferent mechanisms. The most likely candidate in helping to modulate present pain intensity is the totality of past experiences of pain specific to each individual. It is in the prefrontal cortex that this information is integrated into the current sensory signals from the noxious stimulation to produce the conscious experience of pain.

This study is important because it shows that the subjective report of pain intensity is highly correlated to objective activations in the cortex, which not only validates the subjective report of pain intensity but also provides insight into the utility of introspection as a means of assessing a conscious experience. The study is also important because it highlights the role of the prefrontal cortex in assigning pain intensity by integrating present sensory information with past experience. Indeed, previous research has shown that expectations about a stimulus based on previous experience have a marked impact on the current subjective experience of pain. Psychological factors, such as hypnotic manipulation of the subjective experience of pain, have been shown to produce significant changes in the activity of the same cortical regions activated in this study but not the thalamus. This underscores how critically these cerebral cortical regions are involved in the cognitive modulation of the subjective experience of pain, which can be strikingly different among individuals even when they are exposed to the very same noxious stimulation, an experimental proof that what we perceive or what transpires into our personal consciousness is not necessarily what there is factually.

The level of cognitive control that our prefrontal cortex is capable of exerting over the more ancient brain structures not only accounts for some of our differential sensitivity to pleasure and pain, but it affects our entire world view and our behavior. In a 2008 publication, Charles Carver, Sheri Johnson, and Jutta Joormann, from the University of Miami, have argued that low serotonergic function is a marker of a deficit in executive control processes, processes that override or inhibit lower-level influences on behavior. A result is that persons with low levels of the neurotransmitter serotonin in their brain or low brain sensitivity to it have an especially weak coupling between their prefrontal cortex and their limbic brain, hence diminished executive control, and are, therefore, particularly responsive to associative and affective cues of the moment. In the cases that are most familiar, a deficit in executive control yields impulsive action and is associated with externalizing disorders, such as substance abuse and psychopathy. In depressed persons, the same executive control deficit often (though not always) leads to automatic (and in that sense "impulsive") inaction, plus absorption in negative emotions.

Brain neurotransmitters do not act in a vacuum, independently of one another. Instead, it is their constant interaction that yields the world view and behaviors that we exhibit. What further determines whether a person with low serotonergic function behaves in an overtly aggressive and antisocial way rather than a passive-aggressive and depressed fashion is the natural structure of their desire-and-dread (or dopamine) circuitry. If their dopamine circuitry is strongly tilted toward desires, making them highly sensitive to rewards, and if this is coupled with low prefrontal executive control due to low serotonergic function, then there is a high probability for them to exhibit externalizing disorders, including ADHD, mania, substance abuse, and the dark triad consisting of narcissism (characterized by grandiosity, pride, egotism, and lack of empathy), Machiavellianism (characterized by manipulation and exploitation of others, egotism, and an orientation toward achievement, dominance, and a "means justify ends" philosophy), and psychopathy (marked by unemotional callousness, impulsivity, extreme selfishness, remorselessness, and generally antisocial behaviors). If, on the other hand, their desire-and-dread circuitry is more tilted toward dread, making them especially sensitive to punishments, then the low executive control induced by low serotonergic function becomes particularly likely to make them susceptible to internalizing disorders, including anxiety and depression.

Many cognitive psychologists now believe that cognition (broadly conceived) uses two kinds of processes. One process—effortful, top-down, symbolic, and reflective—is used for planning and strategic behavior. This is the domain of the dorsolateral prefrontal cortex along with the rest of the future-oriented desire-and-dread circuitry. Working together, they create and deal with everything in what we called the extrapersonal space, things we do not currently have. The other process—automatic, reflexive, bottom-up, and associationist—is used for acts that are heuristic, skilled, habit-based, or urgent. This is the domain of the pleasure-and-pain circuitry, which deals with everything in the here and now—or what we called the peripersonal space.

There is evidence that these two processing modes learn in different ways, and that the two patterns of learning create parallel and competing paths to action, which require continuous arbitration between them. The two-mode models are also labeled *dual-process models*. They are based on the premise that humans experience reality via two systems. One is a symbolic processor—the rational mind. The other is associative and intuitive, uses shortcuts and heuristics, and functions automatically and quickly. Both systems are always at work, and they jointly determine behavior. The intuitive system almost always gets a first shot at solving a problem. If the task at hand exceeds its abilities, then the slower, reflective mode takes charge of the situation. If it, too, becomes overwhelmed, then the task is handed back to the intuitive system, which may apply the best heuristic solution it has in its arsenal, however ill-fitted it may be to the current problem. The two-mode models essentially explain how the brain deals with immediate versus delayed rewards and punishments. Research on the topic of delay of gratification posits that the brain often must make a choice between a smaller, less desired but immediate and certain reward versus a larger, more desired but uncertain reward later. The two-mode system—a "hot" system (emotional, impulsive, reflexive, and connectionist) and a "cool" system (strategic, flexible, slower, and unemotional)—determines the ability to restrain us when faced with this dilemma and many other contexts. How a person responds to a difficult situation depends on which system presently or generally dominates.

While the behavioral approach and behavioral avoidance systems determine the level of extraversion and neuroticism in someone's personality, cognitive control is key to determining the conscientiousness and agreeableness factors. Indeed, conscientiousness is about exerting effortful control in task-oriented situations, while agreeableness consists in controlling behavior and emotion in people-oriented contexts. This is why psychopaths, who tend to be particularly excited about rewards and especially deficient in executive control, are notoriously irresponsible and unfriendly, although they can exhibit superficial social charm to get what they want.

To appreciate the level of reward sensitivity and the lack of conscientiousness and empathy in criminal psychopaths, consider the case of Bart Whitaker. According to a 2019 article by Peter Van Sant on *CBS News*, by all accounts, the Whitakers were a perfect upper-middle-class family living in a community outside of Houston, Texas. That was all shattered on December 10, 2003, when the family of four was gunned down as they entered their home upon their return from church. Tricia Whitaker, a retired teacher, and her nineteen-year-old son, Kevin, died from their injuries. The dad, Kent, the comptroller of a construction company, survived, as did their son Bart. The investigation into the shooting revealed an intricate plot and a ruthless conspiracy to eliminate three of the four Whitaker family members. The mastermind of the plot? Bart, who hated his family and wanted their money. It turns out that Bart had been lying to his family for years prior to the incident by pretending to attend college while spending away the generous funds his parents kept giving him to cover his education and living expenses. In 2007, he was tried and convicted of hiring the hit man to kill his family and hurt him, too, to cover up the crime. He was given the death penalty. On February 22, 2018, about forty minutes before his scheduled execution time, Bart had his death sentence commuted to life imprisonment without parole by the governor of Texas, in large part due to the influence of his father, who, to the disbelief of some, had fought to save Bart from being executed since he was convicted.

There is evidence that the extreme traits exhibited by psychopaths have biological underpinnings. Based on a 2022 article published by the Nanyang Technological University in Singapore on *SciTechDaily* (a platform dedicated to science and technology news), Olivia Choy, a neuroscientist from this university, working in collaboration with Adrian Raine, from the University of Pennsylvania, and Robert Schug, from California State University, discovered that the ventral striatum, the area of the forebrain responsible for the initiation of motivated action, was 10 percent bigger in psychopathic people compared to a control group of individuals with low or no psychopathic traits. Previous research had shown that psychopaths have an overactive striatum, but the influence of its size on behavior had yet to be confirmed. This new research demonstrated a significant biological difference between people who exhibit psychopathic tendencies and those who do not. While not all people with psychopathic traits end up violating the law, and not all criminals satisfy the criteria for psychopathy, there is a strong association. There is also significant evidence that psychopathy is associated with more aggressive behavior. The study highlights the fact that, in addition to social and environmental influences, it is important to consider that there can be differences in biology, in this case, the size of certain brain structures, between antisocial and neurotypical individuals. Because biological traits, such as the size of one's striatum, can be inherited to a child from a parent, these findings give added support to neurodevelopmental perspectives of psychopathy—the fact that the brains of these offenders do not develop normally throughout childhood and adolescence. Through analyses of the brain scans and results from the interviews of 120 participants to screen for psychopathy, the researchers linked having a larger striatum to an increased need for stimulation through thrills and excitement, a higher likelihood of impulsive behaviors, and a higher probability of going to extreme lengths to seek out rewards, including criminal activities that involve property, sex, and drugs. This points to a highly dopaminergic personality in psychopaths. Previous research has demonstrated that brain scans of incarcerated psychopaths also show deficient connectivity between their prefrontal cortex and limbic brain, hinting at low serotonergic function.

In contrast to psychopaths, conscientious and agreeable individuals are capable of effortful control. Effortful control concerns attentional management (both in terms of sensitivity to new stimuli and in terms of the focusing or persistence of attention during long-lasting tasks) along with inhibitory control (the ability to suppress an approach behavior when doing so is situationally appropriate). The label "effortful" conveys the sense that this is an executive, planned activity, entailing the use of cognitive resources to deter the tendency to react impulsively. The brain's reflexive mode is often characterized as "emotional." This label sometimes seems to imply that emotional experience is subjectively salient in this mode, while sometimes signifying that this mode is very responsive to emotions triggered by situational cues. This system dominates when speed is needed (as when a situation is emotionally charged) and also when processing resources are diminished. That is, it requires relatively little capacity and, thus, can function under suboptimal conditions. It tends to respond to short-term contingencies without consideration for the future or for broader consequences of the action. Thus, although it can be highly adaptive with respect to short-term contingencies, it may be less adaptive when a long-time perspective is more beneficial. The reflexive system is also used when we perform activities that have become routine, such as when we drive a car on "autopilot" while daydreaming at the same time. The reflective (or rational or explicit) system, in contrast, operates mostly consciously, uses logical rules, is verbal and deliberative, and is, therefore, comparatively slow. This rational system provides a more cautious, analytic, planned way of proceeding. It anticipates future conditions, makes decisions based on those predictions, and forms top-down intentions. It is more wide-ranging in its search for relevant information. Because the reflective system requires substantial cognitive capacity, it loses efficiency under high mental load or other conditions that limit cognitive capacity. Planned behavior is not necessarily devoid of emotion. It is clear that planning ahead (as in delay of gratification) implies evaluating the affective consequences of various courses of action. However, an urgent and intense emotion appears likely to compete with, and to potentially short-circuit, the tendency to be deliberate.

Given the differences in their operating characteristics, the brain's reflexive and reflective modes of operation have differing influences on behavior, translating, for the most part, to responding impulsively versus exercising restraint. Each mode of functioning can promote either action or inaction. If enough effortful control capacity is available, the impulsive grabbing of incentives that arises from a sensitive approach system can be restrained. However, effortful control is, in principle, not entirely a matter of restraining approach. Effortful control sometimes means forcing the production of an action that one does not want to take (overriding a reflexive tendency toward inaction). For example, effortful control can lead a sedentary adult to exercise; it can lead a person to stay engaged in a boring task; or it can lead children to look happy when they receive gifts they do not like. Effortful control may be required for a spendthrift to inhibit the tendency to buy things, but it may also be required for a penny-pincher to overcome the tendency not to buy things. Thus, exerting effortful control can move a person either toward restraint or toward action, depending on what reflexive response is being overcome.

We discussed the role of the amygdala in assigning aversive salience to stimuli in our chapter covering the topics of desire and dread. The amygdala is also activated by novelty—for instance, an unfamiliar face (with no emotion) versus a familiar one. The reactivity of the amygdala appears to be a fairly stable trait; children characterized as inhibited at two years of age show hyperreactivity of the amygdala as adults. Low serotonergic function is associated with higher amygdala reactivity, and at the genetic level, the presence of the short allele of the 5-HTTLPR gene in the individual. Despite the fact that the prefrontal cortex is generally linked to higher-order processing and executive control, activation in some prefrontal cortex regions has been consistently linked to automatic and affective processing. For example, some ventral and medial areas of the prefrontal cortex with strong direct connections to subcortical structures, including the amygdala and the striatum, are activated in response to emotion-eliciting stimuli. In contrast, the dorsolateral prefrontal cortex gets activated when people engage in reflective top-down processing, as in reappraising emotional stimuli in order to regulate emotional responses to them. It is also involved in the control of cognition, inhibiting attention to distracting information, and disallowing habitual response tendencies when a task changes. The dorsolateral prefrontal cortex is not directly connected to subcortical areas, though it does connect closely to ventral and medial areas of the prefrontal cortex, which have strong reciprocal connections to subcortical areas. When the reflective dorsolateral prefrontal cortex is not inhibiting the more reflexive ventral and medial areas of the prefrontal cortex, the person is more impulsive. The dorsolateral prefrontal cortex develops in the late teens and early twenties, while the ventral and medial parts of the prefrontal cortex come into their final forms in the midteens. This is the reason why adolescents show more impulsive and risk-taking behaviors than people in their twenties, who seem relatively more mature. Adults with dorsolateral prefrontal cortex damage (such as those afflicted with dementia) have been characterized as resembling children in terms of emotional reactivity and difficulties in regulating emotion.

In their 2008 publication, Charles Carver, Sheri Johnson, and Jutta Joormann propose that serotonergic function influences the balance between the reflective and reflexive processing modes. This is based on previous research showing that increased serotonin levels may decrease the sensitivity of the amygdala to external stimuli, particularly the aversive ones, and increased cognitive control. Research also shows weaker coupling between the amygdalae and the cortical areas in charge of regulating them in people with the short allele of the 5-HTTLPR gene, hence low serotonergic function. This would permit greater excitability of the amygdala and deficits in affect regulation in those individuals. Studies generally indicate that experimentally increasing serotonergic function reduces responsiveness to negative emotional stimuli, decreases aggression, and increases cooperativeness. By contrast, reducing serotonergic function impairs performance on tasks with emotional elements; it also increases expression of aggression and deterioration of cooperativeness, though these effects often occur only among persons who are relatively high in aggressive tendencies by disposition.

There is a substantial basis for believing that ADHD is actually a disorder of the inability to control behavioral impulses. Symptoms include distractibility, difficulty following instructions, disorganization, and talking out of turn. Evidence of actual deficits in attention has generally been lacking. These characteristics likely stem from deficits in executive functions and consequent deficits in motor control. A recent study of children with ADHD found a prospective link from low serotonergic function to antisocial personality disorder nine years later. Lower serotonergic function has long been linked to having a history of fighting and assault, domestic violence, and impulsive aggression more generally, particularly among men. Suicidal behavior has also been linked to serotonin, via serotonin metabolites in cerebrospinal fluid. Although negative affective traits such as neuroticism and hopelessness are involved in the development of suicidal ideation, a large body of research suggests that suicidal acts and completed suicides are reliably tied to impulsivity, particularly impulsive aggression.

We already cited literature relating the presence of the short allele of 5-HTTLPR to a higher likelihood of developing depression in challenging life situations. Persons with depression have difficulties exerting effortful cognitive control. Their attention is drawn to negative emotional cues (perhaps because of their many negative associations in memory), from which they then have trouble disengaging. They are prone to rumination about negativity, but they are also prone to mind wandering rather than task engagement. Those recovered from depression experience recurrence of sensitivity to negative cues when their serotonin levels are artificially lowered. And the fact that serotonin reuptake inhibitors are currently the most effective treatment for depression provides yet another proof that low serotonergic function represents a vulnerability to depression. Women are twice as likely as men to develop internalizing disorders, such as anxiety and depression, while men are twice as likely as women to develop externalizing disorders, such as psychopathy and substance abuse disorders. Although both types of disorders imply low serotonergic function, they result from a differentially built dopamine system, highly sensitive to rewards in men and highly sensitive to punishments in women, perhaps as a result of the developmental influence from the sex hormones testosterone and estrogen. These tendencies presumably stem from the dominant versus submissive roles that men and women have traditionally assumed. All of this taken together appears consistent with the idea that serotonergic function relates to a group of traits with agreeableness and constraint at one end and impulsivity along with either aggressive or depressive tendencies at the other end.

In summary, deficits in cognitive control enhance responsiveness to the cues of the moment, particularly emotional cues. What kind of behavior follows from that responsiveness, however, depends on what the salient cues are. That, in turn, depends partly on the person's emotional and motivational state (or sensitivity). Thus, low serotonergic function and the resulting deficits in effortful control may have divergent effects in people who vary systematically in other ways. When poor executive oversight is combined with moderately high incentive sensitivity (a fairly reactive approach system), the result is impulsiveness. When poor executive oversight is combined with high punishment sensitivity (a fearful avoidance system), the result is inaction and blunted incentive sensitivity. In both cases, the effects of variation in level of incentive versus fearful sensitivity are amplified by the absence of effortful or cognitive override. So, while psychopathy reflects a weak threat system and a strong incentive system, depression works the other way around.

All this makes our innate sensitivity to delayed rewards and punishments, the domain of our brain's dopamine circuitry, a core component of human personality, hence central to the notion of happiness. While our pleasure-and-pain circuitry matters as well because it allows us to enjoy the rewards we currently have and suffer from the mistakes already made, the desire-and-dread circuitry determines everything that lies in the future, both near and distant, all the things that are soon to become our present. It is this circuitry that helps us build mental representations of people and things, which it incorporates into models about how they work together, ultimately forging our worldview. Our dorsolateral prefrontal cortex makes predictions about future events and, using emotions, directs our behavior based on these models. The world can be a rather dark and scary place for the people whose dopamine circuitry is more sensitive to threats than to rewards than average, while it can be ripe with thrills, excitement, and opportunities for those lucky enough to have a desire-and-dread circuitry that reasonably favors rewards over threats. Luckily or unluckily, depending on circumstance, our actual experiences constantly change our brains for the better or for the worse over time. Although the brain is most plastic in youth, it still has some capacity to be reshaped later in life. The caveat is that events of greater emotional magnitude may be required to trigger substantial changes past the age of about thirty. These changes happen through the process of learning, and sometimes through trauma. Bad models lead to bad predictions and costly mistakes. By learning from our mistakes, as well as from our successes, we can modify, destroy, and rebuild our mental models to better fit our reality and make better predictions to guide our behavior going forward.

Thus, rewards and punishments produce learning. There are two main pathways through which the process of learning happens. When Pavlov's dog hears a bell, then sees a sausage, it salivates. If done often enough, the dog will salivate merely on hearing the bell. We say that the bell predicts the sausage, and that is why the dog salivates. This type of learning occurs automatically, without the dog doing anything except being awake. It is called Pavlovian conditioning. Operant conditioning, another basic form of learning, requires the animal's participation. When Thorndike's cat runs around a cage until it happens to press a latch, it can suddenly get out and eat. The food is great, so the cat presses again and again. Operant learning requires the subject's own action; otherwise no reward will come and no learning will occur. If a punishment (such as an electric shock) is delivered following a lever press, then the animal learns to avoid pressing the lever. Pavlovian and operant learning constitute the building blocks for behavioral reactions to rewards and punishments. According to Wolfram Schultz, from the University of Cambridge, both types of learning stem from what is called prediction errors.

To understand prediction errors, we distinguish between a prediction about a future reward and the subsequent reward. Our prediction is based on our existing models of the world. Then we compare the reward with the prediction we made; the reward is better than, equal to, or worse than its prediction. The future behavior will change depending on the experienced difference between the reward and its prediction, the prediction error. If the reward is different from its prediction, a prediction error exists and we should update the prediction and change our behavior. This involves updating our mental models. Specifically, if the reward is better than predicted (positive prediction error), which is what we all want, the prediction becomes better and we will do more of the behavior that resulted in that reward. If the reward is worse than predicted (negative prediction error), which nobody wants, the prediction becomes worse and we will avoid this the next time around. In both cases, our prediction and behavior changes; we are learning. By contrast, if the reward is exactly as predicted, there is no prediction error and we keep our prediction and behavior unchanged; we learn nothing and our mental models remain intact. The intuition behind prediction error learning is that we often learn by making mistakes. Although mistakes are usually poorly regarded, they nevertheless help us to get a task right at the end and obtain a reward. If no further error occurs, the behavior will not change until the next error. This applies to learning for obtaining rewards, as well as it does for learning movements. The whole learning mechanism works because we want positive prediction errors and hate negative prediction errors. This is apparently a mechanism built into the human brain by evolution; it pushes us to always want more and never want less. This is what drives life and progress. It is also what makes us buy a bigger car when our neighbors do so (the neighbors' average car size serves as a reference that is equivalent to a prediction). Even a Buddhist, who hates wanting and craving for material goods and unattainable goals, wants more happiness and rewards rather than less. Thus, the study of reward prediction errors touches the fundamental conditions of life.

Dopamine is not only responsible for helping us to build mental models of the world, but it is also in charge of updating these models based on our reward prediction errors. Dopamine shows a baseline activity level in the brain, called tonic, and occasional bursts of activity in direct response to rewards and punishments, called phasic. While tonic dopamine is responsible for ensuring our normal functioning based on existing models, phasic dopamine updates these models based on new information. More reward than predicted produces a temporary positive burst in dopamine activity above baseline, as much reward as expected induces no phasic dopamine beyond baseline, and less than predicted reward (or a punishment) leads to a negative response, a temporary depression of the dopamine signal compared to baseline. These responses have been observed not only in rodents but also in dopamine neurons in the brains of monkeys and humans. But this is not all. The dopamine response is adjusted for the next preceding reward-predicting stimulus, and ultimately for the first predictive stimulus. The longer the time between the first stimulus and the final reward, the smaller the dopamine response, as subjective reward value becomes lower with greater delays, a phenomenon known as temporal discounting; dopamine responses decrease in specific temporal discounting tests. Otherwise stated, dopamine knows how much a reward may be delayed in time and codes the associated uncertainty of obtaining it by discounting for the time delay! Dopamine neurons show larger responses to risky compared with safe rewards in the low range, in a similar direction to the animal's preferences.

The dopamine neurons of our brains are like "little devils" inside us that drive us to rewards! This becomes even more troubling because of the particular dopamine response characteristics, namely the positive dopamine response (activation) to positive prediction errors; the dopamine activation occurs when we get more reward than predicted. But any reward we receive automatically updates the prediction, and the previously larger-than-predicted reward becomes the norm and no longer triggers a dopamine prediction error surge. The next same reward starts from the higher prediction, and hence induces less or no prediction error response. To continue getting the same prediction error, and thus the same dopamine stimulation, requires getting a bigger reward every time. The little devil not only drives us toward rewards, but it drives us toward ever-increasing rewards. This is what psychologists refer to as the hedonic treadmill. The hedonic treadmill is the reason why it is never enough to get one promotion at work; we soon get used to it and begin to itch for the next one. We buy a bigger house in a more upscale neighborhood only to realize that some neighbors have an even bigger house, and that is what we want next. The dopamine prediction error response underlies our drive for always wanting more reward. This mechanism would explain why we need ever higher rewards and are never satisfied with what we have. We want another car, not only because the neighbors have one, but because we have become accustomed to our current one. Only a better car (or at least a new one) would lead to a dopamine response, and that drives us to buy one.

Dopamine neurons are even more devilish than explained so far. They are at the root of addictions to drugs, food, and gambling. We know, for example, that dopamine mechanisms are overstimulated by cocaine, amphetamine, methamphetamine, nicotine, and alcohol. These substances seem to hijack the neuronal systems that have evolved for processing natural rewards. Only this stimulation is not limited by the sensory receptors that process the environmental information because the drugs act directly on the brain via blood vessels. Also, the drug effects mimic a positive dopamine reward prediction error, as they are not compared against a prediction, and thus induce continuing strong dopamine stimulation on their postsynaptic receptors, whereas the evolving predictions would have prevented such stimulation. The overstimulation resulting from the unfiltered impact and the continuing positive prediction error-like effect is difficult to handle for the neurons, which are not used to it from their evolution, and some brains cannot cope with the overstimulation and become addicted.

On the positive side, the same hedonic treadmill also drives us to ever greater achievements, which drives human progress. Both workaholism (the compulsion or the uncontrollable need to work incessantly) and drug addiction (the compulsion or the uncontrollable need to consume illicit drugs interminably) are the result of a hyperactive dopamine system. A workaholic is addicted to work because work is a proxy for future social and material rewards. A drug addict's behavior is driven by the drug-driven dopamine firing in their brain, creating the mirage of a future reward. So the brain of a workaholic resembles that of a drug addict at the biochemical level.

As the saying goes, "Hard work can beat talent when talent does not work hard." Intelligence and talent will get you far—very far, in fact. Sometimes, the thing that trumps everything else is, however, how much effort you put into something you really want. A workaholic will persist where everyone else has given up. If they also happen to not only be lucky enough to be bestowed with above average talent or intelligence but to benefit from a good hand of fortune as well, then they stand a chance to build the next great American business, write the next best-selling novel, make the next medical breakthrough, solve a mathematical problem never solved before, or make a discovery that changes the way we live for generations. Otherwise stated, a hyperactive dopamine system can be the root cause of fame, fortune, and achievement at one end and mental illness and misery at the other end.

Rather than signaling every reward or punishment as they appear in the environment, phasic dopamine responses represent the crucial term underlying basic, error-driven learning mechanisms for rewards and punishments. Having a neuronal correlate for a positive reward prediction error in our brain may explain why we are striving for ever greater rewards. This behavior is certainly helpful for surviving competition in evolution, and competition is at the root of every single life-improving discovery. But it is a double-edged sword, as it can also generate frustrations and inequalities that endanger both individual happiness and the social fabric. Yet knowledge is power. Having learned about the impact that our brain's circuits for processing both present and future rewards and punishments can have on our well-being, we can set ourselves goals and take actions that are likely to increase positive affect, decrease negative affect, lead to more life satisfaction and, ultimately, greater happiness within the specific range allowed by our biological makeup.

II

What Behavior Can Do: Optimizing Our Subjective Well-Being

5

Cognitive Well-Being: Improving Life Satisfaction

Many thinkers characterize life as a tragedy. Sophocles wrote, "Not to be born surpasses thought and speech. The second best is to have seen the light, and then to go back quickly whence we came." Many behavioral scientists also believe that humans are predominantly dissatisfied and unhappy. Yet an extensive review of existing research on subjective well-being conducted in 1996 by Ed Diener and Carol Diener, from the University of Illinois, suggests otherwise. In fact, most people report a positive level of subjective well-being and say that they are satisfied with domains such as marriage, work, and leisure. A positive hedonic level refers to experiencing positive affect more of the time than negative affect. Subjective well-being is the overall level of happiness a person feels. Positive subjective well-being reflects both a positive evaluation of one's life circumstances, which is what we call cognitive well-being, and positive hedonic level, which represents affective well-being.

For example, Gerald Gurin, Joseph Veroff, and Sheila Feld reported, in 1960, that 89 percent of Americans placed themselves in the "very happy" or "pretty happy" groups; only 11 percent said that they were "not too happy." Even people in disadvantaged groups report, on average, positive well-being, and measurement methods in addition to self-report indicate that most people's affect is primarily pleasant. Cross-national data suggest that there is a positive level of subjective well-being throughout the world, with the possible exception of very poor societies. In 86 percent of the forty-three nations for which nationally representative samples were available at the time of the Diener review, the mean subjective well-being response was above neutral. It turns out that relatively few individuals carry most of the burden of human suffering!

Cognitions tend to be positive worldwide. People think positive thoughts more often than negative thoughts and are more likely to recall positive than negative memories. Most people can also recall positive events from their lives more quickly than negative events. People can use downward comparison to boost their positive affect. There is strong evidence that most people believe that they are better than average on most dimensions. Optimistic cognitions can lead to increased subjective well-being, and the average person appears to be prone to optimism. Because cognition and emotion are so intimately intertwined, the strong evidence for a positive predilection in cognition supports the finding that most people experience predominantly positive affect. It is so amazing to some people that quadriplegics and other people with severe disabilities could be happy that their self-reports are sometimes dismissed as unbelievable. It should be noted, however, that individuals who use wheelchairs are believed by their friends and family to be happy, they can recall more good than bad events in their lives, they are rated as happy by an interviewer, and they report more positive than negative emotions in daily experience-sampling measures.

These findings suggest that the biological hedonic set point is positive for most individuals. The hedonic set point is a baseline affective level toward which we return after positive or negative life events move us away from this baseline. Although the set point varies depending on a person's temperament, for most people it appears to be in the positive range. Ed and Carol Diener speculate that the set point for affect may be positive rather than neutral or negative for several reasons. First, a positive set point gives negative events maximum informational value because they stand out against a positive background. Indeed, a system that is preset to be slightly positive allows threatening events to be noticed quickly. Second, it is important for motivational reasons that people not be in a negative mood most of the time. Approach tendencies must prevail in behavior for people to obtain food, shelter, social support, sex, and so forth. Because positive moods energize approach tendencies, it is desirable that people be, on average, in a positive mood. Human approach tendencies are manifest in the rapid exploration and settlement of new frontiers and in the unremitting invention of new ideas and institutions throughout human history. Thus, not only might humans' large brains and opposable thumbs be responsible for the rapid spread of humanity across the globe, but positive emotions might also be an important factor. Finally, a positive set point may motivate human sociability, drive exploration and creativity, and produce a strong immune response to infections.

Yet most people are not elated most of the time—they are just mildly happy. Average ratings of happiness on a scale of zero to ten are usually a little above seven in the United States. In 2002, Ed Diener and Martin Seligman screened a sample of 222 undergraduate students for happiness. Then they compared the upper 10 percent of consistently very happy people in the sample with average and very unhappy people (those who scored in the lowest 10 percent of the sample). The very happy people were highly social and had stronger romantic and other social relationships than less happy groups. They were more extraverted, more agreeable, and less neurotic. Given these biological traits, it is not surprising that they also scored lower on several psychopathology tests, except for the hypomania score. The very happy group virtually never scored in the clinical range, whereas almost half the individuals in the very unhappy group did so. The researchers found that, for this particular sample of undergraduate students, no variable was sufficient in itself to explain happiness, but good social relations were necessary. Members of the happiest group experienced positive, but not ecstatic, feelings most of the time, and they reported occasional negative moods. This suggests that very happy people do have a well-functioning emotional system that can react appropriately to life events.

While these results are interesting, they report on one-time measurements from a relatively homogenous sample of young undergraduate students with relatively little life experience and similar lifestyles. It would be even more interesting to know if and how major life events affect our long-term well-being by following the same set of randomly chosen individuals over long time periods and recording their self-reported levels of happiness as life events happen to them. In daily life, most people assume that major life events such as marriage or unemployment or becoming paralyzed following a car accident have tremendous effects on happiness. Yet for decades, many researchers claimed quite the opposite. Getting married or divorced, winning the lottery or losing a fortune in a financial crash, getting hired or fired, buying that convertible car or wrecking it—according to these researchers, none of these events should affect the level of subjective well-being for more than a few months because people adapt quickly and inevitably to any life changes.

A major issue with these studies is the fact that they are almost always cross-sectional, meaning they study one sample of people before marriage, divorce, or other life event, and then they sample another set of people after the life event, and they base their conclusions on the aggregate responses of two entirely different sets of individuals. Longitudinal studies, however, have often yielded quite different results. Longitudinal studies follow the very same set of individuals over many years to record their emotional responses to life events as they occur during the study period. Because they follow the same set of individuals before and after the life event, their results should match the reality of these individuals much more closely, and hence be more reliable and telling. Despite this fact, cross-sectional studies have prevailed because they are less time-consuming, logistically easier, and cheaper to conduct.

We already discussed how our temperamental makeup determines a set point for our well-being, a baseline about which our daily emotional state fluctuates. We also discussed how each of us has a genetically set emotional range, centered around our baseline, within which those fluctuations can take place. While people presenting with bipolar disorder have some of the widest possible emotional ranges, those who score very low on both extraversion and neuroticism have extremely narrow ranges. We presented results by Bruce Heady whose 2006 analysis of longitudinal data from the

German Socioeconomic Panel (SOEP) showed that, although subjective well-being changes little over long time periods for most people, small but nontrivial minorities record substantial and apparently permanent upward or downward changes in happiness in response to certain life events. In 2012, Maike Luhmann, from the Freie Universität Berlin, and collaborators wanted to find out whether these results generalized to other samples besides the SOEP. In other words, what is the initial reaction and the rate of adaptation in studies conducted in different cultures, within different time frames, and with different outcome measures? Second, do life events have similar effects on the two main components of subjective well-being, namely, affective and cognitive well-being? With these objectives in mind, they conducted a thorough review of longitudinal data from 188 past publications representing 313 samples and 65,911 individuals. These studies investigated the reaction and adaptation to four family events (marriage, divorce, bereavement, childbirth) and four work events (unemployment, reemployment, retirement, relocation/migration) and were conducted not only in the field of psychology but also in related disciplines such as medicine, sociology, and economics.

Our daily emotions and moods represent our affective well-being and function as an "online" monitoring system of our progress toward our goals and strivings. This system might be highly reactive toward short-term changes of external circumstances, but to retain its informational functionality, it must adapt quickly to long-term changes. Therefore, it can be assumed that for affective well-being, adaptation is functional because adaptation is an essential component of any homeostatic system. Although it might be possible to modify (for instance, decelerate) the rate of adaptation to a certain degree, it is rather unlikely that changing external circumstances will have a long-lasting effect on affective well-being. Changes in cognitive well-being, by contrast, may be less automatic. Cognitive well-being reflects people's life evaluations. For example, income should be an important criterion for this evaluation because making money is a central goal for most people. Similarly, major life events should have measurable and lasting effects on cognitive well-being if they threaten important family- or work-related goals. Hence, the researchers hypothesized that life events have more persistent effects on cognitive well-being than on affective well-being.

They also adopted a working definition of life events according to which life events are time-discrete transitions that mark the beginning or the end of a specific status. A status is a nominal variable with at least two levels. For instance, marital status can be single, married, separated, divorced, or widowed. Occupational status can be employed, unemployed, studying, and so on. The transition from one status to another is a specific life event, for instance, marriage (from single to married), divorce (from married to divorced), job loss (from employed to unemployed), or reemployment (from unemployed to employed). This narrow definition excludes minor life events such as daily hassles (which do not imply a status change) and slow transitions such as puberty (which are not time-discrete). Also, nonevents (such as not finding a marital partner or involuntary childlessness) were not examined within their meta-analysis. Their definition also implies that most life events can be reversed. This phenomenon is common for events such as marriage (through separation) and job loss (through reemployment) and less common for events such as bereavement (through remarriage) and retirement (through reentry into the job market). In their meta-analysis, the researchers also examined how reversing the life event affected adaptation. The adaptation process is initiated by a major life event that causes a physiological or psychological response—for instance, a decrease in subjective well-being. Over time, the responsiveness diminishes and the level of subjective well-being returns to its pre-event level. For negative life events, adaptation is comparable to a recovery trajectory in which normal functioning temporarily gives way to clinical or subclinical psychopathology, usually for a period of at least several months, and then gradually returns to pre-event levels. At this point, it is crucial to point out the potential existence of anticipatory effects. Most major life events are—at least to some extent—controllable and predictable and can therefore be anticipated. This anticipation might cause a specific hedonic reaction even before the event occurred. For instance, if the spouse is terminally ill, the hedonic reaction to bereavement starts long before the spouse actually dies. These so-called *anticipatory or lead effects* can be observed for several months or even years before the occurrence of the event and, therefore, needed to be taken into account in the interpretation of these meta-analytic findings.

How does getting married affect happiness? According to this meta-analysis, the initial reaction to getting married appears to be positive for life satisfaction but not for relationship satisfaction or affective well-being. Over time, both life and relationship satisfaction decline. This does not necessarily mean that getting married makes people unhappier than they were before. Rather, cognitive well-being is higher than usual right before the marriage due to anticipatory effects, and the observed decline reflects a return to premarital levels of subjective well-being. The analysis shows that this "honeymoon effect" is short-lived—adaptation starts quickly, especially if relationship satisfaction is considered. For affective well-being, in contrast, no changes over time were observed. This does not necessarily contradict the assumption that the rate of adaptation is higher for affective well-being than for cognitive well-being. Rather, the weak initial reaction suggests that marriage does not affect affective well-being at all, and consequently, no adaptation is required.

Divorce is typically seen as a negative life event. The findings of this meta-analysis indicate, however, that after a relatively mild negative reaction, subjective well-being increases after divorce. Just as the decline in happiness after marriage does not imply that marriage is inherently negative, this increase in subjective well-being after divorce does not imply that divorce is inherently positive. It is plausible that the level of subjective well-being in the months prior to divorce might be lower than the habitual level, for instance, because people anticipate the divorce and react to it before it occurs. In sum, the findings indicate that the legal act of divorce itself (though not necessarily the whole process) may actually be beneficial for subjective well-being, at least for those who perceive it as a relief from a bad marriage.

Bereavement is usually seen as one of the worst life events and is associated with lasting negative effects on subjective well-being. In this data set, the initial impact of bereavement on subjective well-being was very negative, especially for cognitive well-being. Interestingly, the rate of adaptation was, however, higher than the one observed for divorce. The rate of adaptation was significantly slower in samples with a high proportion of males, suggesting that women adapt faster than men to bereavement. The time period of adaptation is longer than following divorce. The reason why it takes the bereaved so much longer to regain their pre-event levels of subjective well-being is that bereavement is associated with a greater initial shock than divorce. People usually go through a lengthy period of bad marriage and lots of internal struggle before acting on the idea of getting divorced. This gives them plenty of time to anticipate the event. By contrast, the time period between the moment when a spouse gets a stage-four cancer diagnosis and their subsequent death may be much shorter, not to mention the fact that most people will be concentrating on a cure rather than death during that time period. So the eventual loss of life often comes as a shock. The researchers found significant differences among individuals in the reaction to the event but not in the rate of adaptation. The hypothesis that adaptation is faster for affective well-being than for cognitive well-being was not supported for bereavement, although researchers did find that bereavement has stronger and, therefore, more persistent effects on cognitive well-being.

The birth of a child affects its parents' subjective well-being in very diverse ways. Life satisfaction and relationship satisfaction tend to decrease after childbirth. The effects are most pronounced for relationship satisfaction; contrary to life satisfaction, childbirth does not even have an initial positive effect on relationship satisfaction. Due to the continuing decrease in the subsequent years, relationship satisfaction after childbirth is permanently below its prebirth level. This finding shows that the birth of a child is a serious challenge for couples. The long-term effects of childbirth on life satisfaction are also negative, but not quite as severe. Bottom-up theories of subjective well-being posit that global life satisfaction is an aggregate of satisfaction with various life domains. Against this background, this study suggests that the decreased relationship satisfaction has some negative effects on life satisfaction, but these effects are partially compensated by other life domains. Despite these detrimental effects on the cognitive well-being of the parents, the birth of a child is not an entirely negative life event. The effects on affective well-being are small but clearly positive. Although parents tend to be less satisfied after childbirth (for instance, because they have less quality time with their spouses), they *feel* more positive affect in daily life (possibly because the baby is a source of positive affect). In summary, childbirth has very different effects on cognitive well-being and affective well-being. Contrary to the initial hypothesis, however, it is cognitive well-being, not affective well-being, for which adaptation is faster.

Unemployment seems to have differential effects on affective well-being and cognitive well-being. For affective well-being, the initial reaction was, on average, negative but also very diverse across studies, ranging from strong negative to moderately positive effect sizes. Over time, the effect sizes did not change significantly. For cognitive well-being, in contrast, a significant negative initial reaction was followed by an increase in cognitive well-being, suggesting that people adapt to unemployment. However, because the initial reaction was so negative, the pre-event level of cognitive well-being was only reached at approximately three years after the event. Hence, unemployment has very persistent negative effects on cognitive well-being. The hypothesis that the rate of adaptation is higher for affective well-being than for cognitive well-being was not supported for unemployment.

In parallel to unemployment, reemployment has differential effects on cognitive well-being and affective well-being. Affective well-being is not much affected by reemployment. Indeed, the initial reaction is close to neutral, and affective well-being increases relatively little over time. A closer look at the data shows that affective well-being was higher than usual both before and after the event, possibly because reemployment might be anticipated and therefore affect the pre-event scores of affective well-being in a positive direction. Anticipation could also be the mechanism that underlies the somewhat surprising finding that the initial impact of reemployment on cognitive well-being was negative; the actual experience of reemployment might be less positive than anticipated and therefore lead to a short-term decrease of subjective well-being. A similar explanation for this finding might be that in the first months after reemployment, the positive effects of having a new job (such as higher income, feeling useful, etc.) are outweighed by the negative effects (such as less time for leisure and family). In summary, the study revealed a significant difference between affective well-being and cognitive well-being in the rate of adaptation to reemployment, but this difference was contrary to the hypothesis according to which the rate of adaptation should be higher for affective well-being than for cognitive well-being.

Retirement is a typical example of a "neutral" event that comes with costs and benefits. On the one hand, most retirees are probably less stressed and have more time for family, friends, and nonprofessional activities. On the other hand, it is accompanied with reduced income, less structured days, fewer work-related activities, and fewer social contacts. In addition, health problems are more likely in retirees simply because of their age. In the case of early retirement, this event might, in fact, be a direct consequence of reduced health. This meta-analysis shows that the initial reaction to retirement is negative for cognitive well-being but not for affective well-being. This finding might reflect exaggerated expectations toward retirement that are disappointed in the first months. Both affective well-being and cognitive well-being increase after retirement in the following months. The hypothesis that the rate of adaptation is higher for affective well-being than for cognitive well-being was not supported for retirement.

Relocation and migration are stressful events that require people to adjust to new circumstances of their daily lives. The number of studies was too low to draw any final conclusions, and more research on the effects of relocation and migration on subjective well-being is clearly needed. To further complicate things, moving to a whole new country is clearly a far more stressful and life-changing experience than moving to a new city within the same country. So the studies might need to distinguish between these two types of relocation. Interestingly, the results of this meta-analysis suggest that the effects of relocation and migration on subjective well-being are rather positive. Overall, subjective well-being is higher after the event than before the event, especially if cognitive well-being is considered. This effect can be explained in several ways: First, relocation and migration might be genuinely positive experiences that have persistent positive effects on subjective well-being. Second, as everyone who has ever moved will admit, relocating is associated with a lot of work and stress that typically starts well before the actual moving date. Thus, the baseline assessments of subjective well-being might be decreased because of this momentary stress, and the increase in subjective well-being after the event reflects a return to the baseline level. Finally, in the months before relocating, people might overestimate the negative effects of relocation. When this event is less negative than feared, subjective well-being increases.

A central result of this meta-analysis is that the initial hypothesis that adaptation should be faster for affective well-being compared to cognitive well-being is generally not true. Instead, it strongly depends on the event considered. An additional interesting finding is that the effects of life events on cognitive well-being are more consistent across different samples than the effects of life events on affective well-being. A possible explanation for this finding is that affective well-being is much more influenced by other variables such as personality, coping strategies, mood regulation, or social support. This study also shows that life events have the potential to affect our baseline hedonic set point through both anticipatory and post-event effects. We eventually do return to our baseline—at least for the eight life events considered here—but this may take a few years or many more years depending on the event. For instance, while it may take up to three years, on average, to adapt to bereavement, research shows that the birth of a child causes a decline in parental happiness that is not reversed until the parents become empty nesters. So the statement that life events only perturb our happiness set point for a few short months is simply not true. Several years represent a significant portion of the average human lifespan. Consequently, the intuitive view that life events can have a significant impact on our happiness should prevail.

In view of the results of this meta-analysis, individual interventions that change people's activities to increase positive affect and decrease negative affect could be more relevant for affective well-being, whereas public policy interventions that focus on changing people's life circumstances could be more relevant for cognitive well-being. Many researchers classify life events according to their hedonic valence or desirability by distinguishing negative, positive, and neutral events, and propose that negative events should have stronger effects on subjective well-being than positive events. In this meta-analysis, however, desirability does not seem to be a very useful category to examine the differential effects of life events on subjective well-being for two reasons: First, it is not obvious for all events whether they are desirable or undesirable. For instance, the initial reaction to divorce was weaker than the initial reaction to presumably neutral events such as retirement and presumably positive events such as reemployment. Second, the researchers did not find that adaptation is slower for events that are typically considered as undesirable (such as bereavement and unemployment) than for events that are typically considered as desirable (such as marriage and childbirth). As expected, they found that cognitive well-being declines after the positive events and it increases after the negative events. According to the analysis, the rate of decline for the supposedly positive events is, however, not systematically higher than the rate of growth for the supposedly negative events. In conclusion, these findings suggest that events cannot generally be classified as desirable or undesirable. On a cautionary note, this finding is based on a very small sample of life events and needs to be replicated for other positive and negative events.

What if the life events were much more consequential than just getting married, divorced, unemployed, or reemployed? These are all reversible events, after all. How about winning a very sizable prize in a lottery or becoming a quadriplegic as a result of an unfortunate accident? Such a study was done in 1978 by Philip Brickman, from Northwestern University, and his collaborators. The researchers compared a sample of twenty-two major lottery winners with twenty-two controls and also with a group of twenty-nine paralyzed accident victims who had been interviewed previously. For a measure of general happiness, respondents were asked to rate how happy they were now (not at this moment, but at this stage of their life). They were also asked to rate how happy they were before winning (for the lottery group), before the accident (for the victim group), or six months ago (for the control group). Finally, each group was asked to rate how happy they expected to be in a couple of years. All ratings were made on a six-point scale ranging from zero for "not at all" to five for "very much."

While this study has been cited again and again in the literature on happiness as a proof that life events do not affect our subjective well-being much at all, a careful look at the actual results reveals a different story. Lottery winners rated themselves as happier than in the past, even several years after the event, and expected to be even happier in the future. Indeed, they gave a rating of 3.77 to their pre-event happiness, 4.00 to their current happiness, and 4.20 to their expected future happiness. The accident victims felt significantly less happy in the present than in the past, although they rated their future happiness quite optimistically. They rated their pre-accident happiness as 4.41, their current happiness as 2.96, and their expected future happiness as 4.32. These results are then compared to the responses by the controls who gave a rating of 3.32 to their past happiness, 3.82 to their current happiness, and 4.14 to their expected future happiness. It is obvious from these results that lottery winners feel happier currently than in the past and happier than controls, while paralyzed accident victims feel significantly less happy in the present than they used to be and significantly less happy than controls, too. Perhaps people expected lottery winners to experience near-manic levels of ecstasy for the rest of their lives, a state of "nirvana," and perhaps they expected the paralyzed accident victims to feel near-suicidal levels of depression for remainder of their lives, that they took these results to mean that neither winning a large prize in the lottery nor losing the freedom to move around normally following a tragic accident meant anything at all for a person's well-being! The fact that the present levels of happiness of lottery winners and accident victims are not as different from the happiness levels of the control group as one might expect does not erase the fact that they are different nonetheless. It is obvious from these results that both lottery winners and accident victims have adapted to some extent to their new circumstances, and this is a good thing because none of us would want to live with either mania or depression forever. But the level of their happiness set point has changed; it has been adjusted slightly upward for the lottery winners and significantly, though not catastrophically, downward for the accident victims. This study should be cited as proof that very consequential events can permanently reset our baseline happiness within our biological range.

The fact that people's default expectation is a lifetime of unending bliss for lottery winners and a lifetime of inconsolable misery for paralyzed accident victims is due to a mental heuristic called focusing illusion in judgments of life satisfaction. The focusing illusion makes us base our judgments on the single most salient feature of person's life (such as winning the lottery or becoming paralyzed as a result of an accident) to the detriment of everything else that may also be taking place in their life. Lottery winners and paraplegics can have good days and bad days like everyone else because the sun may be shining, the old acquaintance who just passed by them failed to acknowledge them, they completed a work project on time, showed up late for an important appointment, and myriad other events, big and small, that can affect our daily mood. More generally, when a judgment about an entire object or category is made with attention focused on a subset of that category, a focusing illusion is likely to occur, whereby the attended subset is overweighted relative to the unattended subset. In particular, when attention is drawn to the possibility of a change in any significant aspect of life, the perceived effect of this change on well-being is likely to be exaggerated.

To demonstrate this, David Schkade, from the University of Texas at Austin, and Daniel Kahneman, from Princeton University, recruited large samples of students in the Midwestern United States and in Southern California. These students rated satisfaction with life overall, as well as with various aspects of life, for either themselves or someone similar to themselves living in one of the two regions. Self-reported overall life satisfaction was the same in both regions, but participants in both California and the Midwest who rated a similar other participant expected Californians to be more satisfied than Midwesterners. There appears to be a stereotyped perception that people are happier in California. This perception is anchored in the perceived superiority of the California climate and is justified to some extent by the fact that Californians are indeed more satisfied with their climate than are Midwesterners. Nevertheless, contrary to the intuitions of the respondents of this study, and contrary to the intuitions of most people in general, the advantages of life in California were not reflected in differences in the self-reported overall life satisfaction of those who actually live there. The relative advantage of California over the Midwest in terms of its climate looms large when a resident of one region considers the possibility of life in the other. When people answer a question about their own life satisfaction, however, their attention is focused on more central aspects of life. The psychological explanation of the focusing illusion is that it is difficult or impossible to simultaneously allocate appropriate weights to considerations that are at the focus of attention and to considerations that are currently in the background.

Monetary considerations generally loom large in people's evaluations of their past, present, or future life satisfaction, although they have been repeatedly dismissed as unimportant to long-term subjective well-being by many social scientists, typically based on analyses of cross-sectional data. A 2020 longitudinal study by Erik Lindqvist, Robert Östling, and David Cesarini clarified the uncertainty about the direction and persistence of the effects of wealth on subjective well-being. They surveyed a large sample of Swedish lottery players about their psychological well-being and analyzed the data following preregistered procedures. They found that, relative to matched controls, large-prize winners experienced sustained increases in overall life satisfaction that persisted for over a decade and showed no evidence of dissipating with time. Follow-up analyses of domain-specific aspects of life satisfaction clearly implicated financial life satisfaction as an important mediator for the long-run increase in overall life satisfaction. A sustained increase in financial life satisfaction is not easy to reconcile with a common folk wisdom according to which lottery winners tend to squander their wealth through reckless spending. These researchers found little evidence, however, of such behavior in their data. Perhaps this type of behavior, however rarely it may happen, tends to make newspaper headlines and attract attention, which may explain the folk wisdom. And perhaps feeling schadenfreude for the lottery winners who end up miserable makes the countless other players who did not win feel better about themselves. In reality, most lottery winners seem to make the financial benefits of their prize last for many years. This study offers further support that complete adaptation to a significant lottery win almost never happens. Winning a large prize in a lottery seems to permanently adjust a person's baseline or set point for happiness upward within their biological range.

If each person is left free to define well-being as he or she pleases when asked the question "All things considered, how would you say things are these days—would you say that you are very happy, pretty happy, or not too happy?" how can we know if the factors they each consider to be the most salient in order to answer this question are the same? Otherwise stated, are the main environmental causes of human happiness the same or comparable for large numbers of individuals? How can the happiness of one person be compared to that of another? According to Richard Easterlin, professor of economics at the University of Southern California, the essence of the answer, suggested by the answers of large samples of people to questions about the sources of happiness, is this: In most people's lives everywhere, the dominant concerns are making a living, family relationships, and health, and it is these three concerns that primarily determine how happy people feel.

To provide scientific evidence for this idea, social psychologist Hadley Cantril carried out an intensive survey in the early 1960s in fourteen countries with highly diverse cultures and widely different socioeconomic development, asking open-ended questions about what people want out of life. In every country, material circumstances, especially the level of living, are mentioned most often, being named, on average, by three-fourths of the population. Next are family concerns—cited by about half—such as a happy family life and good relations with children and relatives. These are followed by concerns about one's personal or family health, which typically are named by about one-third of the people. After this, and about equal in importance, at around one-fifth of the population, are matters relating to work (such as having a good job) and to personal character (such as emotional stability, personal worth, self-discipline, etc.). Perhaps surprisingly, concerns about broad international or national issues, such as war, political or civil liberty, and social equality, are not often mentioned, being named, on average, by less than one person in twenty. Abrupt changes in these political circumstances can affect people's sense of well-being at the time that they occur, but ordinarily they are taken as given, and it is the things that occupy most people's everyday life and are somewhat within their control that are typically in the forefront of personal concerns. This similarity of environmental causes of happiness across diverse populations stems from the simple fact that most people everywhere spend the majority of their lives doing the same types of things in order to survive and reproduce. This is not to say that the happiness of one individual can be directly compared to that of another, but it does give credence to studies comparing large groups of people.

In analyzing data from the 1994 United States General Social Survey, Easterlin finds that those reporting themselves very happy range from 16 percent in the lowest income class to 44 percent in the highest. By standardizing people's happiness ratings to a scale between zero and four, he further finds that reported happiness shows a positive correlation with income throughout the entire range of incomes, from a low of 1.8 at the lowest incomes to a high of 2.8 at the highest ones. This is carried out by separating people into income ranges and averaging their reported happiness ratings for each income range. It turns out that our intuitive sense that one can achieve higher levels of subjective well-being with higher income is very much true! This is because money can remediate many—though not all—ills in a person's life and open doors that would have remained closed otherwise. How could these facts possibly not impact a person's well-being? Yet this is exactly what many researchers have been asserting for decades. According to them, money brings happiness only to the extent that it lifts a person out of abject poverty into roughly the middle class and makes no difference at all beyond that point. What if the extra money allows a person to set up a dream business or launch a successful political career? What if it allows another to help a family member in need? What if it permits a third person to travel the world and gain experiences they would never have otherwise gained? How about a fourth one who sets up a trust fund dedicated to cancer research using the billions of dollars they earned in their lifetime? Do these scientists really want to assert that none of these accomplishments and experiences—all of which are enabled by money—matter in any way, shape, or form to the well-being of these people?

Data shows that the relationship of happiness with income is logarithmic, meaning that happiness rises rapidly as income is increased from zero up to a point where basic needs are comfortably satisfied, and it continues to rise (forever) as income is increased beyond that point, albeit at increasingly reduced rates. A slowing rate of increase still results in an increase overall, not a plateau! In practice, this is a fortunate thing. Indeed, every single billionaire would be living in a state of constant (and perhaps even worsening) mania if the rate of increase in happiness with income was to be maintained at its initially aggressive levels. Having more money is, thus, one way that we can live within the upper half of our biologically set range for happiness. What money will not do is correct for any shortcomings in temperament that are hard-wired genetically, at least until we reach a point where we can buy genetic interventions. Having money and power can—and will—bring to light any biological defects as well as any talents that poverty and the lack of power had kept hidden. For instance, if a person is naturally prone to addiction and has been using narcotics in the past, having extra money may mean the ability to purchase even more illicit drugs and all the ills that can come with that kind of behavior. All the rock stars and wealthy celebrities who spend their lives getting in and out of drug rehabilitation facilities until they die of an overdose are public testaments to this fact.

Another thing that money will not buy is happiness above and beyond the upper end of a person's biologically set range for subjective well-being. Anthony Bourdain (the celebrity chef and world traveler) and Kate Spade (the world-renowned designer) were two individuals blessed with extremely fortunate life circumstances by any definition but happened to commit suicide in midlife (within months of each other). They are proof that good looks, fame, and fortune are not enough to ease the suffering caused by repeated bouts of depression and anxiety, a biologically hard-wired feature of their brains. Excepting those with extreme scores on certain personality traits, more money does seem to buy more happiness, albeit at increasingly smaller rates. According to Easterlin, in every representative national survey ever done, a significant positive bivariate relationship between individual happiness and income has been found. The relationship holds for household income both adjusted for family size and unadjusted. To discount the positive relationship between happiness and income is to discount the personal testimony of individuals in country after country who mention economic circumstances most frequently as a source of happiness.

Another interesting fact to come from the Hadley Cantril survey conducted in the 1960s is that, when asked to rate their past, present, and future happiness on a scale of zero to ten, in every country and every age group, respondents, on average, rated their prospective happiness higher and their past happiness lower than their present happiness, with only a few trivial exceptions. Many more surveys made since then have confirmed this tendency: People at any point in the life cycle typically think that they will be better off in the future than at present, and that they are better off today than in the past. The 1978 data from Brickman and collaborators on lottery winners and accident victims also showed this tendency among two of the three groups considered, the lottery winners and the controls, the accident victims being the exception. This would, in principle, agree with the data-driven fact that, on average, income (and economic circumstances, more generally) improve substantially up to the retirement ages, at which point it levels off and declines somewhat.

Given that happiness increases with income, people's reported well-being should increase over the life cycle, at least up to the average retirement age, both in reality and as reflected in their ratings of happiness for their past, present, and future. When people are separated into different age cohorts, and then asked to rate their happiness at a given point in time, however, no significant differences can be seen between the average level of present happiness among the cohorts. Easterlin has proposed to explain this paradox by considering not just people's income but also their aspirations at any point in time, as well as over time. Suppose that at the start of the adult life cycle (for instance, in the last year of high school) people in different socioeconomic circumstances have a fairly similar set of aspirations. Those with higher incomes a few years later (say, those who got a college degree) will then be able to better fulfill their aspirations than those with lower incomes (say, those with just a high school degree) and, all other things being equal, will, on average, feel better off. This is the point-of-time positive association between happiness and income. If income rises while material aspirations remain constant, then individuals will experience increasingly higher levels of happiness as they are realizing more of their constant set of aspirations. If income remains constant and aspirations rise, however, then the satisfaction derived from a given level of income would diminish. In reality, material aspirations change over the life cycle roughly in proportion to income. As both income and aspirations rise, they roughly offset each other, resulting in constant levels of happiness over the life cycle as reported by people in different age cohorts. How does one explain the statements on past and prospective welfare? The key is to realize that these are point-of-time responses and are consequently based on the aspirations that people have acquired at that point in time. In other words, people think they were less happy in the past and will be happier in the future because they project current aspirations to be the same throughout the life cycle, while income grows. But since aspirations grow along with income (a direct consequence of the dopamine-induced hedonic treadmill feature of the human brain), experienced happiness is systematically different from projected happiness. This will lead most people to make choices for the future based on false expectations. This, in turn, can lead to disappointment when goals that we so avidly pursued are finally realized. The distinction between decision utility—the perceived satisfaction associated with choice among several different alternatives—and experienced utility—the satisfaction realized from the outcome actually chosen—needs to be taken into account when one is interested in knowing about the consequence of behavior or choice on welfare or happiness.

Here, Easterlin is only expressing in economic terms what we previously discussed in neurological terms; in order to help us achieve our long-term goals, the future-oriented dopamine system of our brains can promise a state of delight that its present-oriented pleasure-and-pain circuitry and its associated neurotransmitters can usually not uphold once the goals are realized. One system excites by dialing up both our mood and activity levels, producing a state of elation, mild anxiety, and tension, while the other one depresses by bringing down worry and anxiety, but also mood and activity, merely inducing a state of satiety, calmness, and contentment. This distinction between the two types of happiness, eudaemonic happiness (or the excitement produced by our imagination from potentially achieving a complex goal with major life impact or obtaining an uncertain, but large future reward) and hedonic happiness (or the pleasure derived from actually experiencing a successful goal or enjoying a smaller, but presently available reward) has perplexed philosophers for millennia and will continue to somewhat disappoint hard-charging individuals, those endowed with a dopaminergic personality, as they realize more and more of their goals. The winners will still enjoy their trophies, only not as much as they had imagined that they would before taking on the challenge, while the losers may be relieved to realize that they do not suffer as much as their brains made them believe, though they will be pained nonetheless.

Selecting the right goals to pursue in life may mean the difference between spending most of the precious little time we are given to live in the upper or lower half of our biologically determined happiness range. If improving our economic circumstances, getting recognition and appreciation from others, and being in good health top the list of people's welfare concerns worldwide, then most major life goals are bound to have one of these three concerns at their core. So what differentiates individual goals is not so much the outcomes that they aim to achieve as it is how they are meant to achieve those outcomes. The goals with the best odds of success are those that are reasonably well matched to the individual's genetic abilities and temperament on the one hand and to their external circumstances on the other. At times, it may be impossible for a person to make further progress toward an important goal because the goal itself is not attainable. Whether this results from a lack of individual skills necessary for realizing the goal or from a depletion of resources and opportunities necessary for its attainment that is due to external circumstances (such as an accident, health problem, or unemployment), facing an unattainable goal creates a problem for a person's quality of life because goal failure has the potential to trigger psychological distress and physical health problems due to chronic stress. If most people are able to maintain a reasonably positive level of life satisfaction by adjusting their goals to be only slightly above what is achievable for them in reality, then perhaps one reason why some people can experience severe melancholia is that they are, for various reasons, unable to reassess and adjust their aims. Carsten Wrosch, from Concordia University in Montreal, Canada, Michael Scheier, from Carnegie Mellon University, and Gregory Miller, from Northwestern University, have proposed a theoretical model postulating that adaptation to unattainable goals requires individuals to *disengage* from the unattainable goal and *reengage* in more feasible goals. In addition, their model assumes that individuals differ widely and reliably in their general goal adjustment capacities. These temperamental differences, in turn, result in major differences in levels of personal happiness if individuals experience unattainable goals.

Goals provide purpose for living, direct individual behavior, and contribute to long-term patterns of successful development. Personal goals can influence quality of life by forming feedback loops, in which a person's perception is compared to a reference value to assess progress toward the goal. If such a comparison process yields a negative discrepancy (for instance, when a person perceives insufficient goal progress), it typically motivates a person to engage in specific behaviors aimed at reducing this discrepancy. The perceived consequences of the ensuing behavioral response are subsequently recompared to the reference value, resulting in a continuous process of goal regulation. Goals thus play an important role in the self-regulation of behavior. In particular, when individuals confront difficulty, their goals can motivate persistent or new behaviors that secure the attainment of desired outcomes and improve the associated quality of life. Being optimistic, believing in one's own competencies, and staying persistent have been shown to be related to subjective well-being and good health. A problem occurs, however, if it is not possible for a person to overcome goal-related problems because there is no behavior that can promote the attainment of a threatened goal. In such circumstances, when a person is confronting an unattainable goal, a likely outcome of persistence is that the person experiences increasingly higher levels of emotional distress. The persistent pursuit of personal goals is only part of adaptive self-regulation, and an equally important part is played by a set of processes that lead to the exact opposite outcome—giving up personal goals when the circumstances warrant it.

One can react in one of two ways to a threatened goal. One way of reacting is continued engagement with the goal. This type of response incorporates the renewal of goal commitment and effort. It should be adaptive if a person has sufficient opportunities to overcome a problem and make further progress toward a threatened goal. If such opportunities are absent or sharply reduced, however, and a person confronts an unattainable goal, the person may need to react with a different type of response. In particular, the latter circumstances may require the person to disengage from the threatened goal and to engage in other, new goals. In such circumstances, goal adjustment processes are likely to facilitate the abandonment of futile endeavors and promote the pursuit of new, meaningful activities.

For instance, research comparing parents whose children had been diagnosed with cancer and parents of physically healthy children documented that goal disengagement capacities were particularly strongly associated with fewer depressive symptoms among parents of children with cancer. Such emotional benefits may occur in the context of demanding life stressors if individuals are able to disengage from goals that have become constrained by the stressor (in this case, their career or leisure goals) and reprioritize time and energy for the most pressing activities (such as caring for their child). Goal disengagement has been scientifically shown to relate to high levels of subjective well-being in people who have developed AIDS, who had handicapped children, who experienced a partnership separation in late midlife, and in women whose biological clock for having their own children had run out. For instance, mothers of autistic children who are relatively happy tend to downgrade the importance of career success in defining their life satisfaction and upgrade the importance of being a good parent, in comparison to mothers who do not have an autistic child. In complementary fashion, cognitive concomitants of failed disengagement, such as rumination and the maintenance of unrealistic intentions, have been shown to be related to periods of distress and depression. For example, new parents who feel the most regret about their past childless lifestyles are also the ones most likely to experience postpartum depression. Goal disengagement is thought to primarily relieve psychological distress by preventing repeated goal failure. It is especially important when one finds oneself in an inescapable situation, such as having to care for a new baby or a handicapped child or being diagnosed with a chronic illness. In addition to providing emotional benefits, if individuals are capable of disengaging from goals that have become constrained by the occurrence of a severe stressor, this process may free resources that can be used to manage the challenging situation more effectively. The primary function of goal reengagement, by contrast, is to keep a person engaged in feasible activities that are meaningful and valuable, which should improve subjective well-being by activating approach tendencies, resulting in both elevated mood and activity levels. Reengaging in more feasible goals after disengaging from unattainable ones is particularly important for young adults who have made the mistake of choosing to pursue a profession for which they have little actual talent or ability. It takes courage to recognize this mistake, give up on the unachievable goal, know oneself sufficiently to choose a more appropriate career path, and then redirect effort toward this new path. Sometimes, individuals adopt goals that are maladaptive or too many in number, which may expend their coping resources and prevent them from effectively addressing pressing life demands (such as an illness or caregiving). In such situations, it may be more useful to maintain a balanced set of goals to prevent conflicts between different goals and protect resources needed for effectively addressing the stressor. Although being able to adjust to unattainable goals is an adaptive personality characteristic, it is not easy to answer the question of when exactly it is an appropriate time to disengage. Indeed, it can be costly to persist in the pursuit of goals that are already out of reach, just as disengaging too early may result in failing to accomplish important life tasks.

Setting the right set of goals for oneself to pursue can thus make all the difference for personal happiness, assuming that external circumstances remain auspicious and the wind keeps blowing in our favor. The right goals to pursue are those that are reasonably well matched to our genetic abilities and temperament on the one hand and our life circumstances of the moment on the other. One must be able to aim a little high in order to make the task of achieving the goal challenging enough but not so high as to miss the mark altogether. The rest is all about dividing the goal into smaller tasks that can be completed within relatively short time periods, managing our time well, being conscientious, and remaining persistent in our effort to bring the goal to completion. Working toward challenging yet realistically achievable goals not only gives structure and meaning to our daily lives, but it also makes it possible for us to have a taste of what psychologist Abraham Maslow called peak experiences, which represent an elated mental state that is most likely induced by the future-oriented dopamine system of our brains. A peak experience is a state of altered consciousness achieved on the way to self-actualization, characterized by a feeling of euphoria. In Maslow's own words, peak experiences are "rare, exciting, oceanic, deeply moving, exhilarating, elevating experiences that generate an advanced form of perceiving reality, and are even mystic and magical in their effect upon the experimenter." This altered state of mind can be reached in the course of doing simple activities, such as during a conversation with a friend that feels so stimulating and absorbing that both participants lose themselves in it, or while immersed in more intense endeavors, such as climbing Mount Everest. Maslow's interviews and questionnaires to gather participants' testimonies of peak experiences revealed that common triggers for such experiences included art, nature, sex, creative work, music, scientific knowledge, and introspection. According to Maslow, often reported emotions in a peak experience include "wonder, awe, reverence, humility, surrender, and even worship before the greatness of the experience," and reality is perceived with "truth, goodness, beauty, wholeness, aliveness, uniqueness, perfection, completion, justice, simplicity, richness, effortlessness, playfulness, self-sufficiency." A peak experience sharpens our senses and involves focused, pleasurable engagement with the activity at hand to the point of losing touch with everything else that surrounds us.

Maslow's concept of *peak experience* is similar to the concept of *flow* introduced by the Hungarian American psychologist Mihaly Csikszentmihalyi. Flow represents a mental state in which a person performing some activity is completely immersed in a feeling of energized focus, full involvement, and enjoyment in the process of engaging in the activity. Csikszentmihalyi came up with this concept when he became fascinated by the sight of artists, especially painters, who got so immersed in their work that they would disregard their need for food, water, and even sleep. As the person fully focuses their attention on the task at hand, they lose awareness of all other things: time, people, distractions, and even basic bodily needs.

Achieving this state of "optimal experience" is very personal and depends on the natural abilities of a given individual. One's capacity and desire to overcome challenges in order to achieve one's ultimate goals leads not only to the optimal experience of flow but also to a sense of overall life satisfaction and happiness. Achieving flow requires active engagement in an activity that must have clear goals with ways to track progress, which establishes structure and direction. The task must provide clear and immediate feedback. This helps to negotiate any changing demands and allows adjustment to performance to maintain the state of flow. Good balance is required between the perceived challenges of the task and one's perceived skills. Confidence in the ability to complete the task is a must. When a person is in a state of flow, they are working to master the activity at hand. According to Csikszentmihalyi, to maintain this mental state one must, over time, seek increasingly greater challenges, otherwise boredom will quickly set in and disrupt the experience of flow. Attempting these new, difficult challenges stretches one's skills. One emerges from such a flow experience with a bit of personal growth and greater feelings of competence and self-efficacy. With increased experiences of flow, people can enjoy personal growth toward increasing levels of complexity. Indeed, people flourish as their achievements grow, and with that comes the development of increasing emotional, cognitive, and social complexity. Csikszentmihalyi warns, however, that "enjoyable activities that produce flow have a potentially negative effect: While they are capable of improving the quality of existence by creating order in the mind, they can become addictive, at which point the self becomes captive of a certain kind of order, and is then unwilling to cope with the ambiguities of life."

The successful pursuit of self-congruent goals that are both challenging and realistically attainable provides us with a way to improve our socioeconomic standing, achieve personal growth, and become all that we can be. This, in turn, allows us to reach a global sense of satisfaction with life, or what we called cognitive well-being. Research reveals that happiness and life satisfaction are similarly available to the young and the old, women and men, blacks and whites, the rich and the working class, as most people tend to be relatively happy at any point in time. Beyond feeling satisfied with one's life, happiness entails experiencing frequent positive affect and infrequent negative affect. For over two millennia, thinkers have offered contrasting ideas about ways to achieve this positive affect balance. They have told us that happiness comes from knowing the truth and from preserving healthy illusions; that it comes from restraint and from purging ourselves of pent-up emotions; that it comes from being with other people and from living in contemplative solitude. Rigorous scientific inquiry is the best way to discern the actual roots of subjective well-being. Let's next look into what science has to say about ways to improve our affective well-being.

6

Affective Well-Being: Experiencing Positive Affect More Often Than Negative Affect

Aristotle regarded happiness as the *summum bonum*, the supreme good. One reason might be the fact that happiness can truly help to extend life. Indeed, a growing body of literature has shown that positive and negative affective states are strongly associated with physical health, mental health, and longevity. For example, in a 1998 analysis of longitudinal data from the Terman Life-Cycle Study, Christopher Peterson, from the University of Michigan, and his collaborators found that the ways in which young men explained bad events predicted health outcomes decades later. The Terman Life-Cycle Study of children with high ability recruited 1,528 respondents in 1922 who were selected on the basis of an intelligence test as being in the top 1 percent of the population. Their development was followed from 1922 to 1995 (with an attrition rate of less than 10 percent) via questionnaires, personal interviews, and various test instruments. Questions were asked about their health, physical and emotional development, school histories, recreational activities, home life, family background, and educational, vocational, and marital histories. Questions were also asked about income, emotional stability, and sociopolitical attitudes. For most of those who had died (about 50 percent of males and 35 percent of females as of 1991), year of death and cause of death were known. In 1936 and 1940, the participants completed open-ended questionnaires about negative life events, which the researchers content-analyzed for explanatory style. Explanatory style is a cognitive personality variable that reflects how people habitually explain the causes of bad events. Among the dimensions of explanatory style are internality ("it is me") versus externality, stability ("it is going to last forever") versus instability, and globality ("it is going to undermine everything") versus specificity. These dimensions capture tendencies toward self-blame, fatalism, and catastrophizing, respectively. Catastrophizing is a cognitive distortion that prompts people to jump to the worst possible conclusion, usually with very limited information or objective reason to despair. The researchers found that the tendency to catastrophize (attributing bad events to global causes) predicted mortality as of 1991, especially among males, and predicted accidental or violent deaths especially well. The reason for this correlation is that emotion-based constructs reflect patterns of coping with negative life events and stresses that can be harmful or beneficial to health. An insightful, positive attitude in dealing with life events, an optimistic explanatory style in contrast to a pessimistic one, can lead to superior ways of responding to adversity, greater feelings of well-being, and perhaps even to longer life.

There are many additional studies that confirm the association between positive emotionality and longer life. In particular, the early-life autobi-ographies and mortality data available for participants in the Nun Study, a longitudinal study of aging and Alzheimer's disease, offer a unique opportunity to investigate the possible association of written emotional expression to longevity. The results from such a study were published in 2001 by Deborah Danner, David Snowdon, and Wallace Friesen, from the University of Kentucky. Participants were members of the School Sisters of Notre Dame religious congregation who, before their retirement, lived and taught in the schools of cities and towns in the Midwestern, Eastern, and Southern United States. In 1991 through 1993, all American School Sisters of Notre Dame born before 1917 were asked to join the Nun Study. Six hundred seventy-eight women agreed to participate in all phases of the study and gave informed written consent to allow access to their archived and active records, participate in annual assessments of cognitive and physical function, and donate their brains at death. What is remarkable about this group of participants is that they all had the same reproductive and marital histories, had similar social activities and support, did not smoke or drink excessive amounts of alcohol, had similar occupations and socioeconomic status, and had comparable access to medical care. This striking similarity in the lifestyles of the participants allows for the elimination of many factors as potential contributors to the study's results.

At the first annual exam, the 678 participants were 75 to 102 years old. Some of these participants had passed away by the time that this particular analysis was conducted. A search of the convents' archives revealed that the Mother Superior of the North American sisters, who resided in Milwaukee, Wisconsin, had sent a letter on September 22, 1930, requesting that each sister write a one-page autobiography. Of the 678 sisters in the Nun Study, 218 took their vows in either the Milwaukee, Wisconsin, or the Baltimore, Maryland, convents and formally joined the religious congregation during 1931 to 1943. Handwritten autobiographies were found for 180 of them. These autobiographies were written sometime between the ages of eighteen and thirty-two. Despite uniformity in the life events that were described, the manner in which the life facts were told in the autobiographies reflected individual style and ranged from simply stating that these life events happened and when they occurred to elaborations of the simple facts that included the emotions experienced by the writer or others involved in the life event. Two coders identified all words in the 180 autobiographies that reflected an emotional experience and classified them as positive, negative, or neutral. Later, a third coder verified each coded word for accuracy and determined the specific type of emotional experience or state referenced by each word. The study found a very strong association between positive emotional content in autobiographies written in early adulthood and longevity six decades later. For example, for every 1 percent increase in the number of positive-emotion sentences there was a 1.4 percent decrease in the mortality rate. Unfortunately, the researchers had no independent measures of temperament, personality, or emotional tendencies for participants, and they could only speculate that individual differences in emotional content in the autobiographies reflect life-long patterns of emotional response to life events in order to explain this association.

It turns out that the long-lived nuns from the School Sisters of Notre Dame religious congregation are not the only people to be more optimistic than pessimistic. In 1988, Shelley Taylor and Jonathon Brown described the general tendency of humans to nurture what they called positive illusions, which consist in unrealistically positive views of the self, exaggerated perceptions of personal control, and unrealistic optimism. They distinguished optimism as an illusion from optimism as a delusion. Illusions are responsive, albeit reluctantly, to reality, whereas delusions are not. Research shows that people in general use more positive words than negative words, whether speaking or writing. In free recall, most people produce positive memories sooner than negative ones. The vast majority of people evaluate themselves positively and, in particular, more positively than they evaluate others. Research on the self-serving bias in causal attribution documents that most individuals are more likely to attribute positive than negative outcomes to the self. One's poor abilities tend to be perceived as common, but one's favored abilities are seen as rare and distinctive. Furthermore, the things that people are not proficient at are perceived as less important than the things that they are proficient at. People substantially overestimate their degree of control over heavily chance-determined events. When people expect to produce a certain outcome and the outcome then occurs, they often overestimate the degree to which they were instrumental in bringing it about. We already reviewed research suggesting that most people believe that the present is better than the past and that the future will be even better. People engage in positive self-deception for a good reason; positive denial can be associated with well-being in the wake of adversity. In difficult circumstances that might be expected to produce depression or lack of motivation, the belief in one's self as a competent, efficacious actor behaving in a world with a generally positive future may be especially helpful in overcoming setbacks, potential blows to self-esteem, and potential erosions in one's view of the future. People are biased toward the positive, and research shows that the only exceptions to this rule are individuals who are anxious or depressed. It appears that the mentally healthy person has the enviable capacity to distort reality in a direction that enhances self-esteem, maintains beliefs in personal efficacy, and promotes an optimistic view of the future. Positive illusions have been tied to reports of happiness. People who have high self-esteem and self-confidence, who report that they have a lot of control in their lives, and who believe that the future will bring them happiness are more likely than people who lack these perceptions to indicate that they are happy at the present.

Being somewhat unrealistically optimistic is thus one way that we can maintain a positive affect balance, and there is good reason to do so. According to Barbara Fredrickson and Marcial Losada, a wide spectrum of empirical evidence documents the adaptive value of positive affect. Beyond their pleasant subjective feel, positive emotions, positive moods, and positive sentiments carry multiple, interrelated benefits. First, these good feelings alter people's mindsets; experiments have shown that induced positive affect widens the scope of attention, broadens behavioral repertoires, and increases intuition and creativity. Second, good feelings alter people's bodily systems; there is research to show that induced positive affect speeds recovery from the cardiovascular aftereffects of negative affect, alters frontal brain asymmetry, and increases immune function. Third, good feelings predict salubrious mental and physical health outcomes; prospective studies have shown that frequent positive affect predicts resilience in the face of adversity, increased happiness, psychological growth, lower levels of cortisol, reduced inflammatory responses to stress, reductions in subsequent-day physical pain, resistance to rhinoviruses, and reductions in stroke. And fourth, perhaps reflecting these effects in combination, good feelings predict how long people live, as demonstrated by the results of the two studies cited at the beginning of this chapter.

Fredrickson explains these benefits with what she calls the broaden-and-build theory of positive emotions, which asserts that positive emotions are evolved psychological adaptations that increased human ancestors' odds of survival and reproduction. The theory holds that, in contrast with the benefits of negative emotions—which are direct and immediately adaptive in life-threatening situations—the benefits of broadened thought-action repertoires emerge over time. Specifically, broadened mindsets carry indirect and long-term adaptive value because broadening builds enduring personal resources, like social connections, coping strategies, and environmental knowledge. As an illustration, consider the link between interest and exploration. Research shows that initially positive attitudes—like interest and curiosity—produce more accurate subsequent knowledge than do initially negative attitudes—like boredom and cynicism. Positivity, by prompting approach and exploration, creates experiential learning opportunities that confirm or correct initial expectations. By contrast, because negativity promotes avoidance, opportunities to correct false impressions are passed by. These findings suggest that positive affect—by broadening exploratory behavior in the moment—over time builds more accurate cognitive maps of what is good and bad in the environment. This greater knowledge becomes a lasting personal resource. Although positive affect is transient, the personal resources accrued across moments of positivity are durable. As these resources accumulate, they function as reserves that can be drawn on to manage future threats and increase odds of survival. So experiences of positive affect, although fleeting, can spark dynamic processes with downstream repercussions for growth and resilience. Whereas traditional perspectives hold that positive affect marks or signals current health and well-being, the broaden-and-build theory goes further to suggest that positive affect also produces future health and well-being. Put differently, because the broaden-and-build effects of positive affect accumulate and compound over time, positivity can transform individuals for the better, making them healthier, more socially integrated, knowledgeable, effective, and resilient.

Fredrickson and Losada predicted that individuals or groups must meet or surpass a specific positivity ratio to flourish based on a nonlinear dynamics model empirically validated by Losada, who studied the interpersonal dynamics of business teams. To flourish means to live within an optimal range of human functioning, one that connotes goodness, growth, and resilience.

This definition builds on path-breaking work that measures mental health in positive terms rather than by the absence of mental illness. Flourishing contrasts not just with pathology but also with languishing, a disorder intermediate along the mental health continuum experienced by people who describe their lives as "hollow" or "empty." Losada's mathematical model predicts that a positivity ratio—the ratio of positive to negative affect—of 2.9, called the Losada line, separates flourishing from languishing. Evidence that there is such a positivity ratio comes from a 1994 study by John Gottman. Gottman and his colleagues observed 73 couples discussing an area of conflict in their relationship. The researchers measured positivity and negativity using two coding schemes: one focused on positive and negative speech acts and another focused on observable positive and negative emotions. Gottman reported that among marriages that last and that both partners find to be satisfying—what might be called flourishing marriages—mean positivity ratios were 5.1 for speech acts and 4.7 for observed emotions. By contrast, among marriages identified as being on cascades toward dissolution—languishing marriages at best—mean positivity ratios were 0.9 for speech acts and 0.7 for observed emotions.

Further evidence corroborating the significance of the 2.9 positivity ratio came from a 2002 study by Robert Schwartz, from the University of Pittsburgh, and his collaborators. They tracked the outcomes of 66 men undergoing treatment for depression and measured positivity ratios before and after treatment. Before treatment, positivity ratios were very low at 0.5. Schwartz and colleagues reported that among patients who showed optimal remission, indexed by both self-report and clinical ratings, mean posttreatment positivity ratios were 4.3. Among those who showed typical remission by the same criteria, mean posttreatment positivity ratios were 2.3. By contrast, among patients who showed no remission whatsoever, mean posttreatment positivity ratios were 0.7. Learning that positivity ratios for flourishing marriages and optimal remission from depression surpassed the Losada line inspired Fredrickson and Losada to test the hypothesis that positivity ratios at or above 2.9 also characterize other samples in flourishing mental health. They recruited 188 students from a Midwestern university to complete an initial survey to identify flourishing mental health and then provide daily reports of experienced positive and negative emotions over twenty-eight days. Results showed that the mean ratio of positive to negative affect was above 2.9 for individuals classified as flourishing and below that threshold for those not flourishing.

We already discussed how tightly related mood and activity seem to be. So one way to reliably lift both our mood and positivity ratio is to increase physical activity through exercise. Numerous studies show a link between physical activity and mood. For example, exercise is a preventive measure for depression and may even have a comparable effect to antidepressants in addressing major depressive disorder (MDD) according to a study by Michael Babyak and collaborators, from Duke University. This study was motivated by the knowledge that aerobic exercise had been prescribed for the treatment of a wide range of medical disorders, including cardiovascular disease, hyperlipidemia, osteoarthritis, fibromyalgia, and diabetes. In addition, exercise was known to have a number of psychological benefits. It had been shown, for example, that physical activity is inversely related to depressive symptoms. In this study, the researchers wanted to know how continued exercise measured up against pharmacotherapy in not only reducing depressive symptoms but also in preventing relapse following recovery. Participants for the study were volunteers aged fifty years and older who met the criteria for MDD. They were randomly assigned to exercise training, antidepressant treatment, or a combination of exercise and antidepressant.

After sixteen weeks of treatment, patients in all three groups exhibited significant reductions in depressive symptoms. Although patients tended to respond more quickly in the medication group, there were no clinically or statistically significant group differences after sixteen weeks. More precisely, 60.4 percent of patients in the exercise group, 65.5 percent in the medication group, and 68.8 percent in the combined group no longer met the criteria for MDD after the four months. Of the original 156 participants, 133 were willing to participate in an additional six-month follow-up. At the end of the four-month intervention, all patients were educated about MDD and were encouraged to continue with some form of treatment on their own, including exercise or medication. Although 64 percent of subjects in the exercise group and 66 percent of subjects in the combination group reported that they continued to exercise, 48 percent of participants in the medication group initiated an exercise program during the six-month follow-up period. The groups differed significantly in the number of subjects using antidepressant medication, with 40 percent of subjects in the combination group, 26 percent in the medication group, and 7 percent in the exercise group reporting antidepressant use during the six-month follow-up period. In addition, twenty-two (or 16 percent) of the participants entered psychotherapy at the end of the four-month intervention. The analysis indicated that the improvements observed after the four-month intervention persisted for at least six months after the termination of treatment. Among patients who had been assessed as being in full remission at the end of the four-month treatment period, participants in the exercise group were less likely to relapse than participants in the two groups receiving medication. Interestingly, combining exercise with medication conferred no additional advantage over either treatment alone. One of the positive psychological benefits of systematic exercise is the development of a sense of personal mastery and positive self-regard, which the researchers believe is likely to play some role in the depression-reducing effects of exercise. It is conceivable that the concurrent use of medication may undermine this benefit by prioritizing an alternative, less self-confirming attribution for one's improved condition. Self-reported participation in exercise during the follow-up period was inversely related to the incidence of depression at ten months. Each fifty-minute increment in exercise per week was associated with a 50 percent decrease in the odds of being classified as depressed. These findings suggest that a modest exercise program (for instance, three times per week with 30 minutes at 70 percent of maximum heart rate reserve each time) is an effective, robust treatment for patients with major depression who are positively inclined to participate in it and that clinical benefits are particularly likely to endure among patients who adopt exercise as a regular, ongoing life activity.

A 2015 analysis by Justin Richards and collaborators based on a large multicountry European dataset representing 11,637 participants from 15 countries found that people felt happier with as little as ten minutes of physical activity per week. Overall, 82.9 percent of the participants reported feeling happy (all the time, very often, or often) in the past month. Significant differences were observed across sociodemographic factors and by country. Italians (72.4 percent) and Germans (73.8 percent) reported the lowest proportions of feeling happy, while the Dutch (91.1 percent) and Irish (90.4 percent) reported the highest proportions of feeling happy. The participants were more likely to report feeling happy if they were male, of younger age groups (meaning below forty years of age), in good health, currently in a relationship, and working. Besides a positive association between educational attainment and feeling happy, those who were still studying were the most likely to report being happy. There was a significant and positive association between physical activity participation level and happiness, with nearly 86 percent of those who were very active reporting feeling happy. A positive dose–response relationship was found for physical activity volume and happiness. Compared to inactive participants, the adjusted odds of being happy was 20 percent higher for people insufficiently active, and this increased to 29 percent and 52 percent higher for those sufficiently and very active, respectively. The results also indicated that the odds of being happy were 2 percent higher for each additional weekly hour of walking and 3 percent higher for each additional hour of vigorous physical activity. Findings from this study indicate that increasing volumes of physical activity are associated with higher levels of happiness. The intensity of physical activity appears to be of minimal importance. Optimal happiness is associated with performing some physical activity at work but not a lot. This concurs with existing evidence for the deleterious effects on mental health of both highly sedentary lifestyles and work that is primarily manual labor. Hence, there appears to be a "happy medium" for vocational physical activity. For best results, it appears best to include a mix of cardio and strength training exercises in one's recreational routine.

Besides getting active to lift our mood when feeling somewhat melancholic, various cognitive methods can be used to control emotions. Counting one's blessings can help to escape the cycle of ruminating on the negative aspects of life. Making it a rule to sit down daily around the same time of day to write down three positive events that happened over the day, no matter how significant or trivial, can redirect attention toward optimism and change our affective tone in the positive direction. Some people go even a bit further by expressing gratitude to the people in their life who may have helped them in some way. If we recall the study of married couples by John Gottman in which he found that flourishing marriages were those characterized by a high positivity ratio, thanking one's significant other for a helpful gesture during the day, even if small, can help push the relationship's positivity ratio upward. Men and women with grateful partners feel more connected to the partner and more satisfied with the romantic relationship. Counting one's blessings and showing gratitude increase happiness by blocking toxic emotions, such as envy, resentment, regret, and sadness. They can also help to reduce aggression and stress, get a better night's sleep, and be more resilient in difficult situations. Counting our blessings and expressing gratitude are two cognitive tools that we have to effortfully change our affective hue from negative to positive.

Meditation is another way to exert cognitive control over our emotions. There are different types of meditation. For instance, mindfulness meditation is based on being mindful, or having an increased awareness and acceptance of living in the present moment. You focus on what you experience during meditation, such as the flow of your breath. You can observe your thoughts and emotions but need to let them pass without judgment. If your mind starts to wander, you need to take notice and slowly return to the object, sensation, or movement that you were focusing on. Spending even a few minutes in meditation can help restore calm and inner peace. It is a simple and inexpensive way to induce relaxation and reduce stress. During meditation, you focus your attention and eliminate the stream of jumbled thoughts that may be crowding your mind and causing stress. This process may result in a sense of calm, peace, and balance that can benefit both emotional well-being and overall health. The benefits of meditation do not end by the end of the session; instead, meditation can help carry oneself more calmly throughout the day.

Another way to escape from intrusive negative thoughts is the use of positive affirmations. A positive affirmation is a phrase or quote you say to yourself in order to combat or challenge overwhelming negative thoughts. For instance, to counter self-deprecating thoughts, one can use affirmations such as "I am proud of myself" or "Today, I choose to be confident." Affirmations help to eliminate the tendency to ruminate or linger on negative feelings while allowing you to acknowledge the things that have prevented your happiness. This, in turn, can increase your self-awareness of what you need and desire, making it easier for you to surround yourself with people and things that support your wants. Whether it is the practice of counting one's blessings, expressing gratitude, meditating, or using positive affirmations, the key to deriving maximum benefit from any of these cognitive tools in the form of lasting happiness is to turn them into habits. All it takes is choosing the cognitive method that best matches one's temperament, making time blocks at regular intervals on our calendar, and holding ourselves accountable for engaging in the activity at the scheduled times. Doing this repeatedly over several weeks should turn the activity into a habit that we simply follow without much thought or effort.

Not being sufficiently attractive physically is a major source of anxiety for many, especially women, given that their socioeconomic standing has traditionally been determined through marriage based primarily on their looks. Since attractive people seem to generally enjoy many social and economic advantages, it is natural to assume that attractive people must be happier than average, and in 2015, Nabanita Gupta, Nancy Etcoff, and Mads Jaeger wanted to find out whether this assumption could be proved scientifically. Indeed, beauty is rewarding and rewarded. Brain imaging studies reveal that brain-reward pathways fire at the sight of attractive strangers' faces, even among infants. Social psychologists have identified a halo effect of physical attractiveness leading to inferences that the attractive are more competent, confident, and socially or professionally skilled than the unattractive. In labor markets, a "beauty premium" and "plainness penalty" are seen: Attractive individuals are more likely to be hired, promoted, and to earn larger salaries than unattractive individuals. Attractive people are more likely to win arguments, persuade others to change their opinions, and be offered assistance. Compared with unattractive adults, they have more dating opportunities and more sexual experience. For all these reasons, one would have to assume that attractive people are happier than other people. But can this assumption be verified scientifically?

To answer this question, Gupta and collaborators used the Wisconsin Longitudinal Survey to test whether attractiveness is significantly linked to psychological well-being and ill-being across the lifespan. The Wisconsin Longitudinal Survey is a long-term study of a broadly representative sample of 10,137 white, non-Hispanic American men and women who graduated from Wisconsin high schools in 1957. Interviews with either the respondents or their parents were conducted six times over a period of more than fifty years between 1957 and 2011 and, for a subsample of the study respondents, four times with a randomly selected sibling. Twelve judges (six men and six women with mean age of 78.5) rated the high-school yearbook pictures of 8,434 participants on an eleven-point rating scale from "not at all" to "extremely" attractive. The researchers used the normed average rating across the twelve judges for their analysis. The participants' measure of body mass index (BMI) was based on self-reports obtained in 1992–1993 and 2003–2005 when subjects were approximately 54 and 65 years old. Finally, they included height in inches.

Their analysis revealed that physical attractiveness can be associated with a statistically significant influence on self-reported well-being and distress. Even when they accounted for education, marriage, widowhood, divorce, illnesses, and income, all known correlates of subjective well-being and depression, the effects remained statistically significant for two out of three measures of attractiveness (facial attractiveness and BMI) on well-being and depression. Facial unattractiveness and elevated BMI may impact well-being and depression through exposure to stigma and discrimination, internalization of stigma, and social disadvantage or ostracism. The subjective well-being measure used for this study focused on feelings of self-confidence, positive self-regard, and agency. The results are consistent with previous studies that found attractive people to be more socially at ease, more assertive, and more likely to think that they are in control of their own lives. Given the social and economic advantages of perceived facial attractiveness when young and of body weight throughout life, it is not surprising that these aspects of appearance may play a role in the development of positive self-regard, self-confidence, and agency. Internalization or transmission of stereotypes can even lead to behaviors that are self-fulfilling prophecies. Today, changing our fate is, however, possible when it comes to beauty. BMI is very responsive to lifestyle changes such as diet and exercise. While, for millennia, physical attractiveness used to be a biologically determined attribute that was nearly impossible to change, modern medicine has made cosmetic and weight-loss procedures not only quite efficacious but also available to the masses. Beauty has become a commodity that money can, literally, buy. In fact, happiness research has established that cosmetic surgery can induce one of the most enduring kind of increase in a person's happiness set point. Despite the negative judgments that cosmetic procedures sometimes elicit, when used in moderation, they can substantially increase a person's self-esteem and overall psychological well-being. Beyond the use of cosmetic interventions, greater societal action aimed at lowering appearance discrimination in the workplace and elsewhere, combined with advocacy of programs to support enhanced body satisfaction and avoidance of messages that decrease body satisfaction, would help to increase subjective well-being for many.

Social relationships—both quantity and quality—affect mental health, health behavior, physical health, and mortality risk, all of which should influence happiness over the course of life. According to Debra Umberson and Jennifer Montez, from the University of Texas at Austin, social relationships have short- and long-term effects on health, for better and for worse, and these effects emerge in childhood and cascade throughout life to foster cumulative advantage or disadvantage in health. Captors use social isolation to torture prisoners of war—to drastic effect. Social isolation of otherwise healthy, well-functioning individuals eventually results in psychological and physical disintegration, and even death. Over the past few decades, social scientists have gone beyond evidence of extreme social deprivation to demonstrate a clear link between social relationships and health in the general population. The most striking evidence comes from prospective studies of mortality across industrialized nations. These studies consistently show that individuals with the lowest level of involvement in social relationships are more likely to die than those with greater involvement. Research provides consistent and compelling evidence linking a low quantity or quality of social ties with a host of conditions, including development and progression of cardiovascular disease, recurrent myocardial infarction, atherosclerosis, poor autonomic regulation, high blood pressure, cancer and delayed cancer recovery, and slower wound healing. Poor quality and low quantity of social ties have also been associated with inflammatory biomarkers and impaired immune function, factors associated with adverse health outcomes and mortality. Marriage is perhaps the most studied social tie. Recent work shows that marital history over the life course shapes a range of health outcomes, including cardiovascular disease, chronic conditions, mobility limitations, self-rated health, and depressive symptoms.

Health behaviors encompass a wide range of personal behaviors that influence health, morbidity, and mortality. According to Umberson and Montez, health behavior explains about 40 percent of premature mortality, as well as substantial morbidity and disability in the United States. Some of these health behaviors—such as exercise, consuming nutritionally balanced diets, and adherence to medical regimens—tend to promote health and prevent illness, while other behaviors—such as smoking, excessive weight gain, drug abuse, and heavy alcohol consumption—tend to undermine health. Many studies provide evidence that social ties influence health behavior. Research across disciplines and populations suggests possible psychosocial mechanisms to explain how social ties promote health. Mechanisms include (but are not limited to): social support, personal control, symbolic meanings and norms, and mental health. *Social support* refers to the emotionally sustaining qualities of relationships (for instance, a sense that one is loved, cared for, and listened to). Hundreds of studies establish that social support benefits mental and physical health. Social support may have indirect effects on health through enhanced mental health, by reducing the impact of stress, or by fostering a sense of meaning and purpose in life. Supportive social ties may trigger physiological sequelae (for instance, reduced blood pressure, heart rate, and stress hormones) that are beneficial to health and minimize unpleasant arousal that instigates risky behavior. *Personal control* refers to individuals' beliefs that they can control their life outcomes through their own actions. Social ties may enhance personal control (perhaps through social support), and, in turn, personal control is advantageous for health habits, mental health, and physical health. Many studies suggest that the *symbolic meaning* of particular social ties and health habits explains why they are linked. For example, meanings attached to marriage and relationships with children may foster a greater sense of responsibility to stay healthy, thus promoting healthier lifestyles. Studies on adolescents often point to the meaning attached to peer groups (for instance, what it takes to be popular) when explaining the influence of peers on alcohol, tobacco, and drug use. *Mental health* is a pivotal mechanism that works in concert with each of the other mechanisms to shape physical health. For example, the emotional support provided by social ties enhances psychological well-being, which, in turn, may reduce the risk of unhealthy behaviors and poor physical health. Moreover, mental health is an important health outcome in and of itself. As the leading cause of disability in both low- and high-income countries, mental disorders account for over 37 percent of the total years of healthy life lost due to disability. Emotionally supportive childhood environments promote healthy development of regulatory systems, including immune, metabolic, and autonomic nervous systems, as well as the hypothalamic-pituitary-adrenal (HPA) axis, with long-term consequences for adult health. Social support in adulthood reduces physiological responses, such as cardiovascular reactivity to both anticipated and existing stressors.

While social relationships are the central source of emotional support for most people, they can also be extremely stressful. For example, marriage is the most salient source of both support and stress for many individuals, and poor marital quality has been associated with compromised immune and endocrine function and depression. Sociological research shows that marital strain erodes physical health and that the negative effect of marital strain on health becomes greater with advancing age. Relationship stress undermines health through behavioral, psychosocial, and physiological pathways. For example, stress in relationships contributes to poor health habits in childhood, adolescence, and adulthood. Stress contributes to psychological distress and physiological arousal (for instance, increased heart rate and blood pressure) that can damage health through cumulative wear and tear on physiological systems, and by leading people of all ages to engage in unhealthy behaviors (for example, excessive food consumption, heavy drinking, and smoking) in an effort to cope with stress and reduce unpleasant arousal. It may seem obvious that strained and conflicted social interactions undermine health, but social ties may have other types of unintended negative effects on health. For example, relationships with risk-taking peers contribute to increased alcohol consumption, and having an obese spouse or friend increases personal obesity risk. Finally, caring for one's social ties may also involve personal health costs. Providing care to a sick or impaired spouse imposes strains that undermine the health of the provider, even to the point of elevating mortality risk for them. Middle-aged adults, particularly women, often experience exceptionally high caregiving demands as they contend with the challenge of simultaneously rearing children, caring for spouses, and looking after aging parents.

According to Umberson and Montez, both quantitative (size and diversity) and qualitative (benefits and costs) aspects of social ties are demographically patterned and socially constructed. Regarding size, women tend to have larger confidant networks than men, as do whites compared with blacks, better-educated adults compared with less-educated, and, to a lesser extent, younger adults. Historically, marriage has conferred more health gains for men than for women. In regard to marriage, men not only experience greater health benefits through the positive lifestyle and health behaviors that often accompany marriage, they also experience fewer costs from spousal caregiving, childrearing, caring for aging parents, and balancing work and family demands. Disparities in the quantity and quality of social ties exist across socioeconomic statuses as well. More educated adults have a larger number of close confidants and may experience less stress in their relationships. For instance, women with a high school degree or less are roughly twice as likely to divorce within ten years of their first marriage compared with women having at least a bachelor's degree. Differential access, benefits, and costs to social ties across sociodemographic groups are, however, not immutable; the right public interventions can change these differentials significantly over time.

Roy Baumeister and Mark Leary have argued that the significant impact of social relationships on mental and physical health is due to human beings having a pervasive drive to form and maintain at least a minimum quantity of lasting, positive, and significant interpersonal relationships. Satisfying this drive involves two criteria: First, there is a need for frequent, affectively pleasant interactions with a few other people, and, second, these interactions must take place in the context of a temporally stable and enduring framework of affective concern for each other's welfare. Interactions with a constantly changing sequence of partners will be less satisfactory than repeated interactions with the same person or persons, and relatedness without frequent contact will also be unsatisfactory. As a result of this basic drive, a lack of belongingness should constitute severe deprivation and cause a variety of ill effects. Furthermore, a great deal of human behavior, emotion, and thought is caused by this fundamental interpersonal motive. The research results cited by Umberson and Montez provide proof for the existence of the need to belong drive among humans. Enhanced relationships and health linkages can be viewed as preventive medicine. While social ties may serve to improve health outcomes for those who develop serious health conditions, they may help prevent these conditions from developing in the first place. Policies that promote and protect social ties should have both short- and long-term payoffs. If social ties foster psychological well-being and better health habits throughout life, then they can add to cumulative advantage in health over time—a worthwhile goal for an aging population. Better health means reduced health care costs, as well as better quality of life for all Americans, regardless of their age. At the individual level, both our psychological and physical well-being might be substantially improved through our personal effort dedicated to enhancing the quality and quantity of our social relationships by nurturing those that are beneficial, severing ties to toxic individuals, and constantly building bonds with new people who might both match and complement us better.

The action of genes on our feelings, thinking, and behavior is pervasive. A 2014 study by Nicholas Christakis and James Fowler found that our friends may be a kind of "functional kin." When it comes to forming relationships with individuals who are neither kin nor mates, we seem to choose those that are more genetically similar to us than not. In fact, the increase in genetic similarity between friends relative to strangers is at the level of fourth cousins! While the people we keep company with can influence our health habits (which can then influence our subjective and physical well-being), we may be choosing them precisely because they share those habits (or the tendency to develop them) with us. Our genes determine our abilities and personality traits, which then influence not only who we choose to spend the majority of our time with, but also the activities we participate in, the jobs that interest us and those that do not, the way that we react to life events, and the environments in which we choose to live—to the extent that we find ourselves in a position to be able make these choices.

It is a general research finding that people who score high on the personality trait of neuroticism are more likely than the general population to experience negative life events and feel the overwhelming burden of negative affect. Those who score low on neuroticism and moderately high on extraversion, on the other hand, may be naturally blessed with a rosy disposition. They tend to experience more positive life events and feel generally happier in life. Stable extraverts may thus be biologically equipped for the highest probability of experiencing psychological well-being, given auspicious life circumstances. Indeed, affective states do not exist in a vacuum; stimuli with hedonic potential are necessary to trigger affective reactions. It is unlikely that even a stable extravert would experience long-term well-being if no pleasurable stimuli were forthcoming from the environment. To the extent that we can exert cognitive control over our emotions, behavior, and external circumstances, it can be helpful to know our natural proclivities so that we can take corrective action when necessary. As unfair as it may sound, this may mean that neurotic introverts may have to work a lot harder than stable extraverts throughout life in order to feel happier. Making life choices that increase happiness—such as pursuing life goals that not only match our temperament and abilities well but also our external circumstances; associating with the right kind of people; striving to be sociable, flexible, open to new experiences, foresightful, self-disciplined, and industrious; being mindful about our exercising, eating, and sleeping habits; and practicing stress management—can benefit all of us, not just those with an overwhelming natural proclivity for negative affect. Luckily, most of us, even those in socially disadvantaged groups, report, on average, positive well-being. Evolution seems to have favored happiness over unhappiness.

REFERENCES

Aristotle. *The Nicomachean Ethics*. London: Penguin Classics, 2004.

Babyak, Michael, James A. Blumenthal, Steve Herman, Parinda Khatri, Murali Doraiswami, Kathleen Moore, Edward Craighead et al. "Exercise Treatment for Major Depression: Maintenance of Therapeutic Benefit at 10 Months." *Psychosomatic Medicine* 62, no. 5 (2000): 633–638.

Baumeister, Roy F., and Mark R. Leary. "The Need to Belong: Desire for Interpersonal Attachments as a Fundamental Human Motivation." *Psychological Bulletin* 117, no. 3 (1995): 497–529.

Berridge, Kent C., Isabel L. Venier, and Terry E. Robinson. "Taste Reactivity Analysis of 6-Hydroxydopamine-Induced Aphagia: Implications for Arousal and Anhedonia Hypotheses of Dopamine Function." *Behavioral Neuroscience* 103, no. 1 (1989): 36–45.

Berridge, Kent C. "Liking and Wanting Food Rewards: Brain Substrates and Roles in Eating Disorders." *Physiology & Behavior* 97, no. 5 (2009): 537–550.

Berridge, Kent C., and Terry E. Robinson. "Liking, Wanting, and the Incentive-Sensitization Theory of Addiction." *American Psychologist* 71, no. 8 (2016): 670–679.

Brickman, Philip, Dan Coates, and Ronnie Janoff-Bulman. "Lottery Winners and Accident Victims: Is Happiness Relative?" *Journal of Personality and Social Psychology* 36, no. 8 (1978): 917–927.

Cantril, Hadley. *The Pattern of Human Concerns*. New Brunswick, NJ: Rutgers University Press, 1965.

Carver, Charles S., Sheri L. Johnson, and Jutta Joormann. "Serotonergic Function, Two-Mode Models of Self-Regulation, and Vulnerability to Depression: What Depression Has in Common with Impulsive Aggression." *Psychological Bulletin* 134, no. 6 (2008): 912–943.

Caspi, Avshalom, Karen Sugden, Terrie E. Moffitt, Alan Taylor, Ian W. Craig, HonaLee Harrington, Joseph McClay et al. "Influence of Life Stress on Depression: Moderation by a Polymorphism in the 5-HTT Gene." *Science* 301, no. 5631 (2003): 386–389.

Chen, Chuansheng, Michael Burton, Ellen Greenberger, and Julia Dmitrieva. "Population Migration and the Variation of Dopamine D4 Receptor (DRD4) Allele Frequencies Around the Globe." *Evolution and Human Behavior* 20, no. 5 (1999): 309–324.

Christakis, Nicholas A., and James H. Fowler. "Friendship and Natural Selection." *Proceedings of the National Academy of Sciences* 111, Suppl. 3 (2014): 10796–10801.

Coghill, Robert C., John G. McHaffie, and Yi-Fen Yen. "Neural Correlates of Interindividual Differences in the Subjective Experience of Pain." *Proceedings of the National Academy of Sciences* 100, no. 14 (2003): 8538–8542.

Costa, Paul T., and Robert R. McCrae. "Influence of Extraversion and Neuroticism on Subjective Well-Being: Happy and Unhappy People." *Journal of Personality and Social Psychology* 38, no. 4 (1980): 668–678.

Csikszentmihalyi, Mihaly. *Flow: The Psychology of Optimal Experience*. United States: Harper Perennial Modern Classics, 2008.

Danner, Deborah D., David A. Snowdon, and Wallace V. Friesen. "Positive Emotions in Early Life and Longevity: Findings from the Nun Study." *Journal of Personality and Social Psychology* 80, no. 5 (2001): 804–813.

Davidson, Richard J. "Well-Being and Affective Style: Neural Substrates and Biobehavioural Correlates." *Philosophical Transactions of the Royal Society of London B* 359 (2004): 1395–1411.

De Martino, Benedetto, Colin F. Camerer, and Ralph Adolphs. "Amygdala Damage Eliminates Monetary Loss Aversion." *Proceedings of the National Academy of Sciences* 107, no. 8 (2010): 3788–3792.

Diener, Ed, and Carol Diener. "Most People Are Happy." *Psychological Science* 7, no. 3 (1996): 181–185.

Diener, Ed, Eunkook M. Suh, Richard E. Lucas, and Heidi L. Smith. "Subjective Well-Being: Three Decades of Progress." *Psychological Bulletin* 125, no. 2 (1999): 276–302.

Diener, Ed, and Martin E. P. Seligman. "Very Happy People." *Psychological Science* 13, no. 1 (2002): 81–84.

Easterlin, Richard A. "Income and Happiness: Towards a Unified Theory." *The Economic Journal* 111, no. 473 (2008): 465–484.

Ebstein, Richard P., Olga Novick, Roberto Umansky, Beatrice Priel, Yamima Osher, Darren Blaine, Estelle R. Bennett et al. "Dopamine D4 Receptor (D4DR) Exon III Polymorphism Associated with the Human Personality Trait of Novelty Seeking." *Nature Genetics* 12, no. 1 (1996): 78–80.

Faure, Alexis, Sheila M. Reynolds, Jocelyn M. Richard, and Kent C. Berridge. "Mesolimbic Dopamine in Desire and Dread: Enabling Motivation to Be Generated by Localized Glutamate Disruptions in Nucleus Accumbens."

Journal of Neuroscience 28, no. 28 (2008): 7184–7192.

Fredrickson, Barbara L., and Marcial F. Losada. "Positive Affect and the Complex Dynamics of Human Flourishing." *American Psychologist* 60, no. 7 (2005): 678–686.

Gilbert, Daniel. *Stumbling on Happiness*. New York: Vintage Books, 2005.

Gottman, John M. *What Predicts Divorce? The Relationship Between Marital Processes and Marital Outcomes*. Hillsdale, NJ: Erlbaum, 1994.

Gozzi, Marta, Erica M. Dashow, Audrey Thurm, Susan E. Swedo, and Caroline F. Zink. "Effects of Oxytocin and Vasopressin on Preferential Brain Responses to Negative Social Feedback." *Neuropsychopharmacology* 42 (2017): 1409–1419.

Gray, J.A. *Neural Systems, Emotion, and Personality*. In J. Madden, IV (Ed.), *Neurobiology of Learning, Emotion, and Affect* (pp. 273–306). New York: Raven Press, 1991.

Gupta, Nabanita Datta, Nancy L. Etcoff, and Mads M. Jaeger. "Beauty in Mind: The Effects of Physical Attractiveness on Psychological Well-Being and Distress." *Journal of Happiness Studies* 17, no. 3 (2015): 1313–1325.

Gurin, Gerald, Joseph Veroff, and Sheila Feld. *Americans View Their Mental Health*. New York: Basic Books, 1960.

Headey, Bruce. "Happiness: Revising Set Point Theory and Dynamic Equilibrium Theory to Account for Long Term Change." *DIW Berlin Discussion Papers* 607 (2006): 1–18.

Herzog, Jan, Julia Reiff, Paul Krack, Karsten Witt, Bettina Schrader, Dieter Müller, and Günther Deuschl. "Manic Episode with Psychotic Symptoms

Induced by Subthalamic Nucleus Stimulation in a Patient with Parkinson's Disease." *Movement Disorders* 18, no. 11 (2003): 1382–1384.

Kant, Immanuel. *Critique of Pure Reason*. London: Penguin Classics, 2008.

Kapur, Shitij, Romina Mizrahi, and Ming Li. "From Dopamine to Salience to Psychosis—Linking Biology, Pharmacology, and Phenomenology of Psychosis." *Schizophrenia Research* 79, no. 1 (2005): 59–68.

Lieberman, Daniel Z., and Michael E. Long. *The Molecule of More: How a Single Chemical in Your Brain Drives Love, Sex, and Creativity—and Will Determine the Fate of the Human Race*. Dallas: BenBella Books, 2018.

Lieberman, Matthew D., and Naomi I. Eisenberger. "Pains and Pleasures of Social Life." *Science* 323, no. 5916 (2009): 890–891.

Lindqvist, Erik, Robert Östling, and David Cesarini. "Long-Run Effects of Lottery Wealth on Psychological Well-being." *The Review of Economic Studies* 87, no. 6 (2020): 2703–2726.

Luhmann, Maike, Wilhelm Hofmann, Michael Eid, and Richard E. Lucas. "Subjective Well-Being and Adaptation to Life Events: A Meta-Analysis." *Journal of Personality and Social Psychology* 102, no. 3 (2012): 592–615.

Lykken, David, and Auke Tellegen. "Happiness Is a Stochastic Phenomenon." *Psychological Science* 7, no. 3 (1996): 186–189.

Maslow, Abraham H. *Religions, Values, and Peak Experiences*. London: Penguin Books, 1964.

McCrae, Robert R., and Paul T. Costa. "Adding Liebe und Arbeit: The Full Five-Factor Model and Well-Being." *Personality and Social Psychology Bulletin* 17, no. 2 (1991): 227–232.

McCrae, Robert R., and Paul T. Costa. "The Stability of Personality: Observations and Evaluations." *Current Directions in Psychological Science* 3, no. 6 (1994): 173–175.

Nanyang Technological University. "Scientists Have Established a Key Biological Difference Between Psychopaths and Normal People." *SciTechDaily*, June 1, 2022.

Pavlov, Ivan P. *Conditioned Reflexes*. London: Oxford University Press, 1927.

Peterson, Christopher, Martin E.P. Seligman, Karen H. Yurko, Leslie R. Martin, and Howard S. Friedman. "Catastrophizing and Untimely Death." *Psychological Science* 9, no. 2 (1998): 127–130.

Plato. *Protagoras*. United States: Hackett Publishing Company, Inc., 1992.

Richards, Justin, Xiaoxiao Jiang, Paul Kelly, Josephine Chau, Adrian Bauman, and Ding Ding. "Don't Worry, Be Happy: Cross-Sectional Associations Between Physical Activity and Happiness in 15 European Countries." *BMC Public Health* 15 (2015): Article 53.

Salvatore, Jessica E., Sara Larsson Lönn, Jan Sundquist, Kristina Sundquist, and Kenneth S. Kendler. "Genetics, the Rearing Environment, and the Intergenerational Transmission of Divorce: A Swedish National Adoption Study." *Psychological Science* 29, no. 3 (2018): 370–378.

Sample, Ian. "Scientists Find Genetic Mutation That Makes Woman Feel No Pain." *The Guardian*, March 27, 2019, at 8:01 p.m. EDT.

Schkade, David A., and Daniel Kahneman. "Does Living in California Make People Happy? A Focusing Illusion in Judgments of Life Satisfaction." *Psychological Science* 9, no. 5 (1998): 340–346.

Schultz, Wolfram. "Dopamine Reward Prediction Error Coding." *Dialogues in Clinical Neuroscience* 18, no.1 (2016): 23–32.

Schwartz, Robert M., Charles F. Reynolds III, Michael E. Thase, Ellen Frank, Amy L. Fasiczka, and David A.F. Haaga. "Optimal and Normal Affect Balance in Psychotherapy of Major Depression: Evaluation of the Balanced States of Mind Model." *Behavioural and Cognitive Psychotherapy* 30, no.4 (2002): 439–450.

Sophocles. *Oedipus at Colonus* (R. Fitzgerald Trans.). In D. Greene & R. Lattimore (Eds.), *The Complete Greek Tragedies* (Vol. II). Chicago: University of Chicago Press, 1959.

Taylor, Shelley E., and Jonathon D. Brown. "Illusion and Well-Being: A Social Psychological Perspective on Mental Health." *Psychological Bulletin* 103, no.2 (1988): 193–210.

Tellegen, Auke, David T. Lykken, Thomas J. Bouchard, Jr, Kimerly J. Wilcox, Nancy L. Segal, and Stephen Rich. "Personality Similarity in Twins Reared Apart and Together." *Journal of Personality and Social Psychology* 54, no. 6 (1988): 1031–1039.

Terman, Lewis M., and Melita H. Oden. *The Gifted Child Grows Up: Twenty-Five Years' Follow-Up of a Superior Group.* United States: Stanford University Press, 1947.

Thorndike, Edward L. *Animal Intelligence: Experimental* Studies. New York: MacMillan Publishers Ltd., 1911.

Tomarken, Andrew J., Richard J. Davidson, Robert E. Wheeler, and Robert C. Doss. "Individual Differences in Anterior Brain Asymmetry and Fundamental Dimensions of Emotion." *Journal of Personality and Social Psychology* 62, no. 4 (1992): 676–687.

Umberson, Debra, and Jennifer Karas Montez. "Social Relationships and Health: A Flashpoint for Health Policy." *Journal of Health and Social Behavior* 51, Suppl. (2010): S54–S66.

Van Sant, Peter. "Texas Man Targeted by a Hitman Fights for the Life of the Person Who Ordered the Murder—His Son." *CBS News*, January 5, 2019, at 11:12 p.m. EDT.

Wrosch, Carsten, Michael F. Scheier, and Gregory E. Miller. "Goal Adjustment Capacities, Subjective Well-being, and Physical Health." *Social and Personality Psychology Compass* 7, no. 12 (2013): 847–860.

www.ingramcontent.com/pod-product-compliance
Lightning Source LLC
Chambersburg PA
CBHW051716090426
42738CB00010B/1939